Practical Statistics for Biologists Workbooks

AN INTRODUCTION TO BASIC STATISTICS FOR BIOLOGISTS USING R

About The Authors: Dr Colin D. MacLeod graduated from the University of Glasgow with an honours degree in Zoology in 1994. He then spent a number of years outside of the official academic environment, working as, amongst other things, a professional juggler and magician to fund a research project conducting the first ever study of habitat preferences in a member of the genus *Mesoplodon*, a group of whales about which almost nothing was known at the time. He obtained a masters degree in marine and fisheries science from the University of Aberdeen in 1998 and completed a Ph.D. on the ecology of North Atlantic beaked whales in 2005, using techniques ranging from habitat modelling to stable isotope analysis. Since then he has spent time working as either a teaching or research fellow at the University of Aberdeen and has taught Geographic Information Systems (GIS) at the University of Aberdeen, the University of Bangor (as a guest lecturer) and elsewhere. He has been at the forefront of the use of habitat and species distribution modelling as a tool for studying and conserving cetaceans and other marine organisms, and he has co-authored over forty scientific papers on subjects as diverse as beaked whales, skuas, bats, lynx, climate change and testes mass allometry, many of which required the use of complex statistical analysis. In 2011, he created *Pictish Beast Publications* to publish a series of books, such as this one, introducing life scientists to key practical skills, and *GIS In Ecology* to provide training and advice on the use of GIS and spatial statistics in marine biology and ecology.

Dr Ross MacLeod graduated from the University of Glasgow with an honours degree in Zoology in 1999 and went on to study risk trade-off behaviours in birds for his doctoral degree at the Edward Grey Institute of Field Ornithology, University of Oxford, graduating in 2004. Since then, he has focused on behavioural ecology and biodiversity conservation research around the world, and has led projects in the UK, Bolivia and Peru. After working as a NERC-funded post-doctoral researcher at the University of St Andrews, he moved to the Institute of Biodiversity & Animal Health (IBAHCM), University of Glasgow as a Royal Society of Edinburgh Research Fellow, investigating how the impacts of environmental change on biodiversity can be predicted from knowledge of animals behavioural decisions and examining how biodiversity conservation can be delivered through sustainable development and rainforest regeneration. In 2018, he was appointed as a Lecturer in Behavioural Ecology at Liverpool John Moores University, with a research focus on forecasting future population and ecosystem impacts of environmental change. Throughout his career he has been involved in developing new ecological survey techniques and skills-based teaching approaches in the fields of biodiversity measurement, environmental monitoring, statistics and, more recently, GIS.

Cover Image: The oak woodland study area used to investigate factors that influence the breeding success of hole-nesting birds, such as great tits (right), as measured by variables like the clutch size (left), the number of chicks that hatched (middle) and the number that fledge, at the University of Glasgow's SCENE field station on the shores of Loch Lomond in Scotland.

PSLS

Practical Statistics for Biologists Workbooks

AN INTRODUCTION TO BASIC STATISTICS FOR BIOLOGISTS USING R

Colin D. MacLeod and Ross MacLeod

Pictish Beast
Publications

ISBN – 978-1-909832-07-7
Published by Pictish Beast Publications, Glasgow, UK.
Printed in the United Kingdom.
First Edition: 2020.

Trademarks

All terms mentioned in this book that are known to be trademarks or service marks have been appropriately capitalised. The author(s) and the publisher cannot attest to the accuracy of this information. The use of a term in this book should not be regarded as affecting the validity of any trademark or service mark. In addition, the use of a trademark should not be taken to indicate that the owner of that trademark endorses the contents of this book in any way, or that the author(s) and/or publisher endorse a particular brand or product.

Warning And Disclaimer

Every effort has been made to make this book as complete and as accurate as possible, but no warranty or fitness is implied. The information provided is on an 'as is' basis and is provided as examples for training purposes only. The author(s) and the publisher shall have neither liability nor responsibility to any persons or entity with respect to any loss or damages arising from the information contained in this book.

'Don't Panic!'

From the cover of *The Hitchhiker's Guide To The Galaxy*

This book is dedicated to those biologists who wish to learn how to use R for statistical analysis without panicking too much.

Table of Contents

Preface

This is the first book in the *Practical Statistics for Biologists Workbooks* series. This series has been created to fill the gap between learning about statistical theory and learning how to actually use statistics in a biologically meaningful way in your research. Thus, these workbooks focus on developing the practical skills and knowledge needed to carry out statistical analyses in R, rather than providing information about statistical theory. This is because there are already plenty of excellent books available that explain statistical theory, but far fewer that help biologists learn the whole process required to systematically carry out statistical analysis in R from start to finish using an approach that can be immediately applied to their own data.

Traditionally, statistics have been taught to biologists using a knowledge-based approach. In this approach, individual components of statistical analysis, including its theoretical background, are taught as more or less separate entities. Each person is then expected to work out for themselves how to integrate these components so this knowledge can be applied to their own work. In contrast, these workbooks use a task-oriented learning (TOL) approach. Rather than focussing solely on knowledge acquisition, this approach focuses on teaching complex tasks, such as statistical analysis, through their practical implementation. This means that instead of simply being shown how to do the individual components of data analysis, you are provided with all the steps you need to do to complete a specific analytical task from start to finish. For example, when learning how to create a frequency distribution histogram (see Exercise 2.1), you are not just provided with instructions for creating this type of graph, but also with details of how to import your data set into R, how to check that it as been imported properly before you create your graph from it, and how to present your the final frequency distribution histogram in a thesis or manuscript. This TOL approach allows you to start applying the workflows for the specific tasks provided in this workbook to your own data almost immediately without first having to learn a large amount of statistical theory, as well as every possible command you can use to implement such data analysis in R. Of course, this doesn't mean it replaces the need to learn about statistical theory or to gain a more detailed understanding of what can be achieved by using R. Instead, it provides the impetus to persist with such learning. Thus, task-oriented learning provides a quick-start approach on which you can build a more detailed understanding of biological statistics.

--- Chapter One ---

Introduction

The aim of this workbook is to introduce biologists to the practical elements of statistical analysis using R statistical software. This means it is primarily aimed at undergraduate and postgraduate students who either wish to teach themselves how to do statistics in R or who are taking their first courses in how to use R. However, it will be just as useful for more experienced biologists who currently use other statistical software packages, but wish to learn how to use R and to quickly come to grips with the practical aspects of using it to analyse biological data.

This workbook uses the same task-oriented learning (TOL) approach found in other books in the *PSLS* series, such as *GIS For Biologists: A Practical Introduction For Undergraduates.* The TOL Approach helps you learn how to carry out the types of tasks biologists need to be able to do on a regular basis in a practical and meaningful way without getting too tangled up in learning about the underlying theoretical basis for them. This workbook, therefore, does not aim to provide you with information about statistical theory (there are already plenty of very good books available on this subject, of which we would recommend the ones by Alain Zuur – see *www.highstat.com/books.htm* for more details). Instead, it focuses on providing practical experience and advice about how to carry out the types of basic data processing and statistical analysis biologists use on a daily basis.

This practical experience and advice comes in the form of a series of exercises which you can work through to learn how to complete specific data analysis tasks. This might be something simple, such as importing data into R and checking it for errors, or something more complicated, like running a linear regression analysis. No matter what, for each task, you are provided with all the steps you need to complete it, starting with getting your data into R and finishing with how to present the results of your analysis to others. These exercises use a workflow approach based around flow diagrams to help you understand exactly what you need to do at each step in the process, and where you are using the same basic steps to complete different tasks. This allows you to see how more complicated tasks can be carried out by connecting together simpler individual steps.

The exercises in this workbook are divided into five groups. These are: 1. Basic data processing tasks, such as importing data into R, error-checking them, subsetting them, joining data from different data sets together and summarising them (Chapter Three); 2. Creating graphs from biological data to look for and show patterns within them (Chapter Four); 3. Assessing and transforming the distribution of biological data (Chapter Five); 4. Comparing data from different groups or samples using statistical analysis (Chapter Six); and 5. Using correlations and regressions to analyse biological data (Chapter Seven). Together, these represent most of the key tasks that biologists need to be able to do to start analysing their data in a practical and meaningful way.

R was selected as the basis for the instructions provided in this book because it is free to download and because it is widely used by biologists around the world. While it is a command-driven package, which some people initially find off-putting, it is relatively simple to use, and with the right type of instructions, it is relatively easy for anyone to successfully start using it for statistical analysis in a short space of time. However, while the exact instructions will vary for other statistical software, the same basic steps outlined in the boxes on the left hand side of the flow diagrams are required to complete the tasks outlined in this workbook in almost all of them. Thus, it is relatively easy to adapt the instructions provided here for use with other analytical software.

The exercises start with the first task you need to be able to do to analyse your data in R, which is importing them into the software (Exercise 1.1), before moving on to more advanced tasks, such as calculating summary statistics (Exercise 1.5), creating bar graphs (Exercise 2.2), carrying out a t-test (Exercise 4.1), and conducting linear regression (Exercise 5.2). This allows you to build up your analysis skills in a logical order as you work through this book one chapter at a time. However, the instructions provided in each chapter are sufficiently complete that you can work through the exercises in each one on their own without having to refer back to the contents of previous chapters. This means if your main aim from reading this book is to work out how to do a specific task, such as making a graph or comparing data from different groups, you can go straight to the relevant chapter for that task and find out how to complete it. If you do this, however, it is worth taking the time at a later date to work through the earlier chapters too, as these will help widen your skills base as well as giving you a better understanding of what you are doing in each individual step within more advanced analyses.

When you start using R you may find it rather frustrating, particularly if you are not already familiar with using command-driven software packages. This is because, rather than picking actions from a menu as you would do with a graphic-user interface, you need to enter each command as a line of text. To make things more complicated, the text needs to be entered in a very precise manner, including using exactly the same uppercase and lowercase letters provided in the instructions for each exercise. This means that any typos you make will cause your commands not to work, and R will not necessarily provide you with suggestions or indications as to what has gone wrong. Since the aim of this workbook is to help you get up and running with biological data analysis in R as quickly as possible, rather than testing your typing skills, you will find a text document called R_CODE_BASIC_STATS_ WORKBOOK.DOC in the compressed folder containing all the data required to complete these exercises (instructions for downloading this folder can be found at the start of Chapter Three). This document provides a copy of all the R code used in the flow diagrams that you will find at the start of each exercise. You can, therefore, choose to simply copy and paste the required code into R from this document without having to worry about making mistakes while typing it. Once you are familiar with how to complete a specific task, you can then work out how to modify this basic code to allow you to do other, related tasks, and you will be given the opportunity to do this as part of each exercise. To help with this, the commands provided in the above document have been colour-coded. This not only makes it easier for you to work out what each different part does, it also makes it easier to work out which bits need be modified to make them do something different. If you wish to learn more about how to create R code to do specific tasks for yourself, we recommend reading *Getting Started with R: An Introduction for Biologists* by Andrew P. Beckerman, Dylan Z. Childs and Owen L. Petchey.

How The Exercises in This Workbook Are Structured:

The exercises in this workbook all follow a standard structure that has been specifically developed to help you to understand what you need to do to complete a specific task in R, to gain experience in doing it, and to help you work out how to apply it to your own data. First, you are provided with a brief introduction to the task itself, and to the structure that your data need to have to be able to complete it. Next, you will find a flow diagram with all the information you need to work through an example of the task using a specific data set. Once you have worked through this initial example, you will find details of how you can modify the commands you used to produce different results, as well as examples of such modifications that you can work through. This will help you gain a deeper understanding of how you can adapt a specific workflow to your own data. At the end, you will find a final example that you can use to test the knowledge you have gained from the exercise. This approach of providing detailed step-by-step workflows, along with examples of increasing complexity for you to work through by entering and modifying variations on a specific set of commands, means that you can use these exercises to rapidly and efficiently increase your data analysis skills, as well as providing a resource you can refer back to any time you wish to refresh your knowledge of how to do a specific task.

Why Are Some Instructions And Steps Repeated In Different Exercises?

As you work through the chapters in this workbook, you will quickly notice that there are some instructions and steps that are repeated in many different exercises. If you are not already familiar with the task-oriented learning (TOL) approach used in this book, you may think this repetition is unnecessary. It is not, and it does, in fact, perform a number of important functions that will help you master the use of R for statistical analysis. First and foremost, it reminds you that there are certain key steps which you need to do each and every time you wish to do an analysis in R. These include steps like setting your working directory, importing a data set and checking that it has been loaded into R correctly. By repeating them in each individual exercise, it not only helps you to become familiar with these basic, but important, steps, it also serves to reinforce the importance of including them in every workflow that you carry out. Secondly, by including the instructions for the same steps in multiple exercises, it enables you work through a specific task, such as doing a t-test, from start to finish. This means you can concentrate on learning all the steps you need to do to complete that task without becoming distracted by having to refer to other sections of the book. Finally, by including the same steps in the flow diagrams for different exercises, it helps you see how you can create workflows for more complex tasks by building up steps from simpler ones. For example, the instructions for completing a t-test in Exercise 4.1 include steps that require you to import a data set into R (from Exercise 1.1), check that it has been imported correctly (from Exercise 1.2), create a box plot (from Exercise 2.4) and assess whether or not the data being analysed have a normal distribution (from Exercise 3.1) as well as the instructions for conducting the t-test itself. This makes it much easier to understand how you can create your own custom workflows for relatively complex tasks not included in this workbook by building up the instructions for simpler steps (see Appendix I for more details on how to do this).

NOTE: As with many things in life, there may be more than one way to do the processes required to complete the exercises outlined in this workbook. The instructions presented here will work for the associated data sets, and this means they should also work in most other circumstances. However, if you find an alternative way to do them which works for you, or if you have someone who can show you how to do them in another way, feel free to do them differently.

--- Chapter Two ---

What You Need To Know To Get Started With R

What Is R?

R is an open source, and so freely available, analytical software package that is widely used by biologists for carrying out data processing and statistical analysis. In fact, it has become so widely used that knowing how to use R can now be considered a critical skill for almost all biologists. However, unlike most other widely used software packages, R does not have a true graphic user interface (GUI) that allows you to simply click on menus and/or buttons to carry out specific tasks. Instead, it is primarily command-driven. This means that you need to enter lines of code in order to get it to do what you want it to do. Many biologists find this rather daunting, and at first it can be, but you should not allow this to put you off using it. This is because R is an amazingly powerful tool that, if you learn to use it properly, can make your life as a biologist so much easier. In addition, once you start using it, you will find that R is not as difficult, or as complicated, to use as it may at first seem, and the effort you put into learning it will be repaid many times over by the benefits of being able to use it to process and analyse your data more quickly and more efficiently than you could ever hope to do without it.

Where Can I Get R From?

As an open source software package, R can be freely downloaded from the R Project website at *www.r-project.org*. This workbook assumes that you will be using version 3.6.1, 3.6.2 or 3.6.3 (the three most recent versions at the time of writing), but the instructions provided in it should work for almost any recent and future versions of R.

How Can I Use R?

There are two common ways that biologists use R. These are using the native R user interface, which will be referred to in this workbook as RGUI, and by accessing it using a third-party user interface. The RGUI option is the one that automatically appears when you open R, and it looks like this:

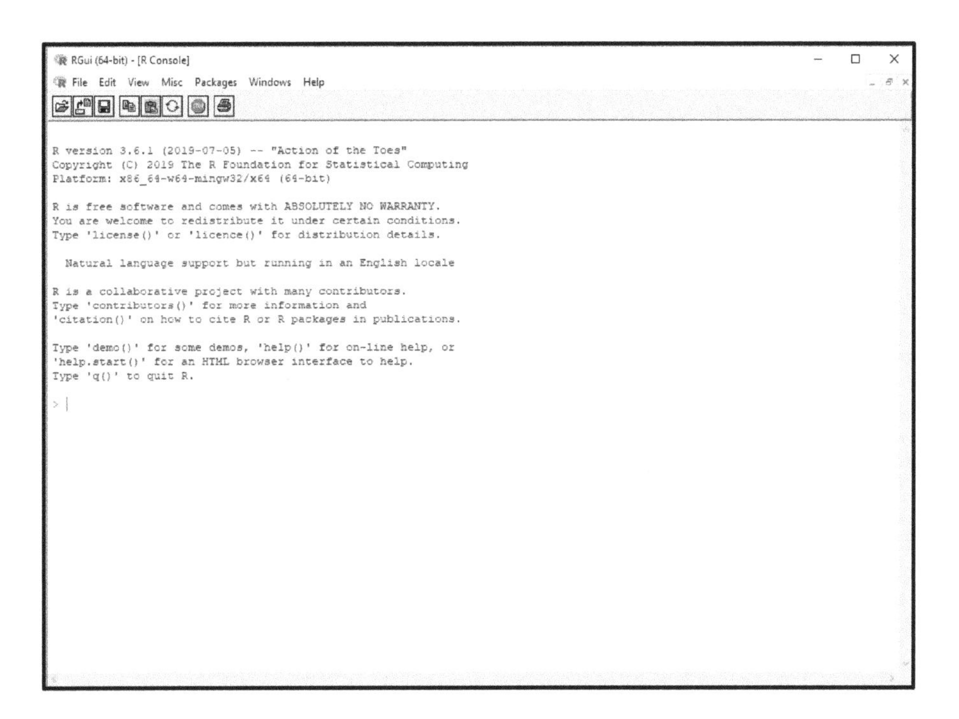

This window (which, when you first open it, contains text detailing of the version of R you are using and various other bits of information) is known as the R CONSOLE window. To use the RGUI, you simply type the R code you wish to run on the command line of this window (the line that starts with a > symbol) and press ENTER on your keyboard. To save your work, or to reload a previous work session, you can select the appropriate option from the menus along the top of the user interface (see below for more details). As and when you call them up, other windows, such as DATA VIEWER windows (which allow you to view data tables that you have within R), R GRAPHCS windows (which will display graphs and other graphics generated by the R commands you enter into the R CONSOLE window) and an EDITOR window, will appear in the main RGUI window alongside the R CONSOLE window.

Of the third-party user interfaces available for R, the one that is the most widely used by biologists is called RStudio. The main reason that many biologists prefer to use RStudio rather than the native RGUI is because it makes it easier to follow exactly what you are doing, to edit the commands you have used in the past to adapt them for use with new data sets and to archive them in a meaningful way so you have a record of exactly what you have done (see Appendix II). As a result, we recommend that you use RStudio to access R, rather than RGUI, both for completing the exercises in this workbook and for your own analyses. However, regardless of how you access R, you will use the same commands to complete the same tasks. Any points of difference between using these two interfaces for the exercises in this workbook are highlighted within the instructions provided. This means that, as far as this workbook is concerned, it does not matter which interface you choose to use.

Unlike the native RGUI, RStudio needs to be downloaded and installed separately. However, it is also freely available and you can download it from *www.rstudio.com*. Once you have installed RStudio, it will automatically connect to R (as long as you have already installed R on your computer) and it will be ready to use. When you first open RStudio, its user interface should look like this:

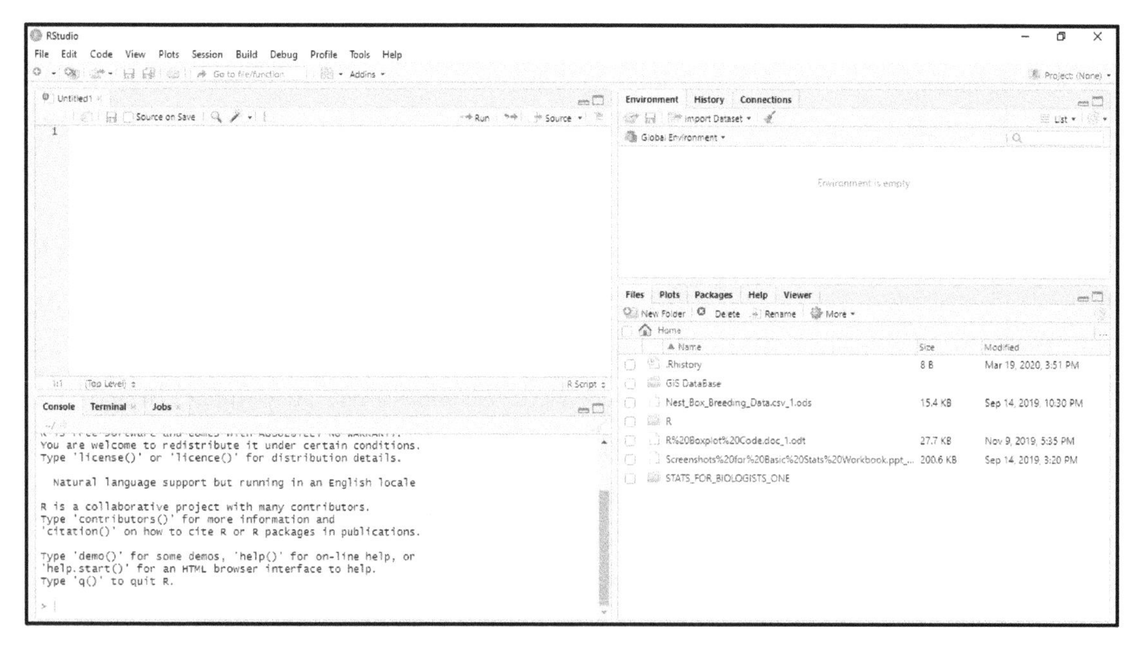

If your version of RStudio does not look like this, click on VIEW on the main menu bar, and select PANES> SHOW ALL PANES. This will make sure that all the required windows are set to display. If the upper left hand window is still not visible after you have done this, click on FILE on the main menu bar and select NEW FILE> R SCRIPT. This should make the missing window appear.

The RStudio user interface is divided into four windows or panes. These are: 1. The SCRIPT EDITOR window (the upper left hand window); 2. The R CONSOLE window (the lower left hand window); 3. The ENVIRONMENT/HISTORY window (the upper right hand window); 4. A window in the lower right hand corner that includes tabs for FILES (which allows you to browse files on your computer – this means the list of files and folders that appears in this tab will be unique to your computer and will not match those in the image on page 8), PLOTS (which displays any graphs or other graphics created by commands you run in the R CONSOLE window), PACKAGES (which allows you to access additional packages – see below for information on what an R package is) and HELP (where you can get help if you get stuck with anything while using R).

As with the native RGUI, you can enter the R code you need to run a specific command directly in the R CONSOLE window in RStudio. However, the main advantage of using RStudio is that you can also use the SCRIPT EDITOR window to enter, edit, prepare, annotate and store any R commands you might wish to use before you run them in the R CONSOLE window. This makes it a much more flexible way to enter commands into R. In order to run a command from the SCRIPT EDITOR window, you need to transfer it to the R CONSOLE window. To do this, you simply select the appropriate command (or a block of commands) in the SCRIPT EDITOR window and click on the RUN button at the top of it.

What Special Terms Are Used When Describing Things In R?

For the most part, the terms that are used when completing data analysis tasks in R are the same as those you would use in any other statistical package. However, there are eight terms which are more specific to R that are useful to understand in order to be able to use it more easily. These are:

1. **Code:** In R, Code refers to the string of text and characters that you need to enter to make it do something, such as importing a data set, creating a graph or running a specific statistical test. A piece of Code is made up of a number of different components,

including Commands, Arguments, Additional Arguments and Objects (see below). In this workbook, R code will always be highlighted using the `Courier New` font (this is the default font for the R CONSOLE window in both RGUI and RStudio).

2. **Commands:** In R, a Command is the name given to a piece of Code that allows you to do a specific function. For example, the `setwd` command allows you to set the WORKING DIRECTORY for a specific analysis project in R (see below), while the `read.table` command allows you to import a data set into R (see Exercise 1.1). The block of Code for a complex task may consist of the number of different Commands, each of which needs to start on a new line.

3. **Arguments:** In R, an Argument is a term you must include in a Command in order for it to work. For example, for the `setwd` command, you need to include an argument that tells it the address of the folder you wish to set as the WORKING DIRECTORY (see below). If you don't include all the required Arguments in the Code for a Command, it will not work. This means that Arguments can be thought of as essential settings you need to specify to make a specific Command work.

4. **Additional Arguments:** Additional Arguments are similar to Arguments, in that they represent settings you can enter for a Command. However, unlike Arguments, they are optional. For example, in the `hist` Command used to make a frequency distribution histogram in R, you can include the Additional Argument `col` to set the colour that will be used for the bars on your graph (see Exercise 2.1). If you don't include this Additional Argument, the Command will still work, but the frequency distribution histogram created by this Command will use the default colours. This means that Additional Arguments can be thought of as optional settings you can include to give you more control over what a Command does.

5. **Objects**: An Object is something created by a Command in R that stores a specific piece of information. Most of the time this will be a table of data, but R Objects can also be created to store things like graphs and the outputs of statistical tests. As such, you can think of R Objects as being equivalent to the tables used by and outputs produced with other software packages. The only difference is that Objects are not saved separately from your R project and can only be accessed through R. If you set a Command to create

a new Object, it will be saved in your analysis project. If you don't, while the results will be shown in your R CONSOLE window, they will not be saved. Within R Code, the name of the Object that will be created by a specific Command is linked to it by either an equals sign (=) or by an arrow created by using a less than sign (<) followed immediately by a hyphen (-) so it looks like this <-. This latter option will be used in this workbook to link each Object to the Command used to create it.

6. **Scripts:** A Script is a collection of R Commands that allows you to do a more complex task from start to finish. They are often stored in a specific file type called an R Script file (which has the extension .R) that can be imported into R, although you can also save them as other file formats, such as text files (.txt). In other analytical software, such as SPSS, Script files are known by other names, such as Syntax files.

7. **Libraries:** A Library is a block of pre-existing Code that tells R how to process a data set when certain Commands are entered into it. This Code runs in the background each time you use one of these Commands and all you will see is the results it returns. This means that in order to be able to use a specific Command from a specific Library, the Library itself has to first be loaded into R. For most basic Commands, the Libraries required for them to work will be loaded into R by default. However, for other Commands, you need to manually load their associated Library each time you wish to use them. This can be done using the `library` Command. If you cannot get a specific Command to work, you should check that its associated Library has been loaded into your current R project (see page 17 for details of how to do this).

8. **Packages:** A Package is a collection of Libraries that allow you to do a specific set of tasks in R. Before you can load a Library from a specific Package and use the associated Commands, you need to install the Package itself. This can be done using the `install.packages` Command. If you cannot load a specific Library that you wish to use into R you should check that the Package which contains it has been installed in your current version of R (see page 17 for details of how to do this).

How Do I Get Started With Using R?

The very first thing you should do before you start any data analysis project in R is to create a new and dedicated folder for it on your computer. You will use this folder to store all the data you are going to use as part of your analysis project, as well as all the outputs that are generated during it and all the code you used to run commands within it. This folder will be what is known as the WORKING DIRECTORY for your project. It is good practice to create a new WORKING DIRECTORY folder for each project you wish to carry out in R. This is because it prevents data, commands and results of different projects getting mixed up. Once you have created a WORKING DIRECTORY folder, you need to copy or transfer all the data you are going to use for that specific project into it.

Next, you need to copy the address for this folder to the clipboard of your operating system. To do this on a computer running a Windows operating system, open Windows Explorer and navigate to your newly created WORKING DIRECTORY folder. Once you have opened it, click on the folder icon at the left hand end of the ADDRESS BAR at the top of the Windows Explorer window to reveal the full address of this folder. You can now copy this address by selecting it and pressing the CTRL and C keys on your keyboard at the same time. **NOTE:** You will need to modify folder addresses before you can use them in R. Specifically, Windows uses backslashes (which look like this \) as the separators between the parts of the addresses. However, R uses slashes (which look like this /). This means that you will need to replace all the backslashes (\) in the address of your WORKING DIRECTORY folder with slashes (/) before R will recognise it. For example, a Windows address that reads C:\USERS\DOCUMENTS\STATISTICS would need to be modified for use in R so that it reads C:/USERS/DOCUMENTS/STATISTICS.

To copy the address of your newly created WORKING DIRECTORY folder on a computer running a Mac operating system, open Finder and navigate to the location of your WORKING DIRECTORY folder. Select this folder and press the CMD and I keys on your keyboard at the same time. This will open the GET INFO window where you will find the address for this folder alongside WHERE. You can then select it and copy it using the CMD and C keys on your keyboard at the same time. **NOTE:** You may find that you have to modify Mac OS folder addresses before you can use them in R. Specifically, you may need to drop the part that says MAC HD at the start and replace the folder separators between the parts of the addresses with slashes (which look like this /). For example, a Mac OS address that reads MAC HD▶USERS▶ADMIN▶DOCUMENTS▶STATISTICS would need to be modified for use in R so that it reads /USERS/ADMIN/ DOCUMENTS/STATISTICS.

Once you have created your WORKING DIRECTORY folder and copied its address, you can open your preferred user interface (either the native RGUI or RStudio). As soon as it opens, you should save a copy of the file where you are going to enter your R commands in this folder. This file is known as your WORKSPACE file. It will have the extension .RDATA and it will contain all the R objects you will create during your project, such as the tables of data you import and the results of statistical analysis you carry out. In RGUI, you can do this by clicking on the FILE menu at the top of the main RGUI window and selecting SAVE AS. This will allow you to save your project as an R WORKSPACE file under a specific name in the WORKING DIRECTORY folder you have created for your specific project. In RStudio, you can save your WORKSPACE file in your WORKING DIRECTORY folder by clicking on the SESSION menu at the top of the main RStudio window and selecting SAVE WORKSPACE AS.

If you are using RStudio, you can also save the contents of your SCRIPT EDITOR window as an R SCRIPT file (which has the extension .R). This file will contain all the code that you have used in your project meaning that if you wish to re-run it, or run a modified version of it, you do not need to type it in again. To save the contents of your SCRIPT EDITOR window as an R SCRIPT file, click on the FILE menu at the top of the main RStudio window and select SAVE AS. This will allow you to save it using a specific name in your WORKING DIRECTORY folder.

After you have saved the contents of your WORKSPACE and the contents of your SCRIPT EDITOR window (if you are using RStudio), the next thing, you need to do is to clear any data that are currently held in the temporary memory of R itself. This prevents you from accidentally using the wrong data set when you start a new analysis project. You can do this by using the following command:

```
rm(list=ls())
```

If you are using the native RGUI, you can simply type this code after the command prompt at the bottom of the R CONSOLE window (it looks like this: >). If you are using RStudio, you can type this command into the SCRIPT EDITOR window (the upper left hand window) and then run it in the R CONSOLE window (the lower left hand window) by selecting it and then clicking on the RUN button at the top of this window.

Next, you need to tell R where the WORKING DIRECTORY you are going to use for your analysis is located on your computer. This is done using the `setwd` command. This command should be followed by the address of the WORKING DIRECTORY folder you have created for your project (and into which you have already saved or transferred all the data you are going to use for your analysis - see above). For example, if your WORKING DIRECTORY folder has the address C:\STATS_FOR_BIOLOGISTS_ONE, this command would read:

```
setwd("C:/STATS_FOR_BIOLOGISTS_ONE")
```

If you are using RGUI, you can set your WORKING DIRECTORY by typing `setwd("` after the command prompt in the R CONSOLE window. Next, paste or type the address of your WORKING DIRECTORY folder after this command (remembering to modify the folder separators in this address as outlined on pages 12 and 13 so R can recognise it) and then type a second quotation mark followed by closing bracket, like this `")`. Finally, press the ENTER key on your keyboard to run this command. Alternatively, you can click on the FILE menu at the top of the main RGUI window and select CHANGE DIR.

If you are using RStudio, you can set your WORKING DIRECTORY by typing setw`d("` into the SCRIPT EDITOR window. Next, paste or type the address of your WORKING DIRECTORY folder after this command (remembering to modify the folder separators in this address as outlined on pages 12 and 13 so R can recognise it) and then type a second quotation mark followed by a closing bracket, like this `")`. Finally, select the whole command then click on the RUN button at the top of the SCRIPT EDITOR window to run it in the R CONSOLE window. Alternatively, you can click on the SESSION menu at the top of the main RStudio window and select SET WORKING DIRECTORY> CHOOSE DIRECTORY.

To check that you have set R to use the correct WORKING DIRECTORY, you can use the `getwd()` command. When you run this command, it should return the same address as the folder you wish to use as your WORKING DIRECTORY.

Once you have set your WORKING DIRECTORY, you are ready to start entering the commands you need to process and/or analyse your data. This may include importing your data into R (Exercises 1.1), dividing them into subsets (Exercise 1.3), creating summary statistics from them (Exercise 1.5), creating graphs from them (Exercise 2.1 to 2.5), assessing whether your data are normally distributed (Exercise 3.1), and running the appropriate statistical analyses to help you explore the research questions you wish to answer (Exercises 4.1 to 5.2). In all cases, when entering your commands, remember that if you are using the native RGUI user interface you can simply paste or type the required R code after the command prompt at the bottom of the R CONSOLE window and then press the ENTER key on your keyboard. If you are using RStudio, you can paste or type the required R code into the SCRIPT EDITOR window (the upper left hand window) before selecting it and then clicking on the RUN button at the top of it to run this code in the R CONSOLE window.

If you wish to load a project you have been working on into RGUI, you can click on the FILE menu and select LOAD to load the contents of a specific WORKSPACE file, including all the objects previously created in it, into R (as long as you remembered to save it the last time you stopped working on it). If you wish to load the contents of a WORKSPACE file into RStudio, click on the SESSION menu at the top of the main window and select LOAD WORKSPACE. If you are using RStudio and you also wish to load an R SCRIPT file containing R code you previously saved from the SCRIPT EDITOR window, click on the FILE menu and select OPEN FILE. **NOTE:** If you re-load an R SCRIPT file into RStudio without also loading the associated WORKSPACE file, you will need to re-run all the R commands you have previously run (such as the commands to import a specific data set) to re-create the required R objects before you can run any new commands based on them.

What Do I Do If I Get Stuck When Using R?

There will be times when using R that you will find yourself getting stuck. When this happens, the most important thing to remember is 'Don't Panic!'. It is completely normal to get stuck when using R and it happens to everyone at some point or other. In fact, when trying to do something new in R, you should always assume that it will take you an average of three attempts to do it successfully (the first time to do it wrong, the second time to work out why it went wrong the first time, and third time to do it right). As a result, any time you manage to do something in less that three attempts should be considered a bonus. Of course, one of the main advantages of R is that once you get a specific set of commands working, you can save them and use them repeatedly without having to trouble-shoot them all over again. In addition, you can share the R code you have created with others to allow them to repeat the same analysis without having to build the same code for themselves from scratch. However, this does require that you keep a good record of exactly what code you have used and what it does (see Appendix II for instructions on how to do this). When you get stuck trying to use a particular command or piece of R code, the situation can generally be resolved using one of the following five solutions:

1. **Check For Typos (Including The Mis-Use Of Capital Letters):** Typos in file names, R object names, variable names, commands and the arguments you can include in them are the number one cause of problems you will encounter with R. As a result, this is the first thing you should check for when anything goes wrong or doesn't work properly.

2. **Check You Are Using The Correct Files, Data Sets, R Objects And Variables:** If you are copying an R command from someone else or from your own R scripts, it is very easy to forget to change one or more of the settings, meaning that a particular command ends up trying to use the wrong information. Therefore, this is the second thing you should check when a command doesn't give you the results you were expecting.

3. **Check The Help Files And/Or Documents For Commands You Are Using To Ensure You Are Using Them Correctly:** Sometimes, you may find that a particular command simply will not be able to do what you wish it to do or that you need to change the settings you are using for it. As a result, if you find that a particular command does not work, and you don't find any typos in it and you are certain that it is being applied to

the correct data, you should carefully read through the help files and/or the documentation for the command to ensure you are using it correctly. You can access the help files for a particular command in R by entering the word 'help' followed by the name of the command in brackets. For example, to get help with a command called `read.table`, you would enter the code `help(read.table)`. Alternatively, you can enter a question mark (?) followed by the command you are looking for help with, like this: `?read.table`. Both of these options will bring up the help file for that command which will provide detailed advice about how you can use it. You can find similar information in the documentation for each command, which can be found at *www.rdocumentation.org*. This information can be located using an internet search engine by entering the name of the command followed by the words *R Documentation.*

4. **Check That All The Required Packages And Libraries Have Been Installed In Your Version Of R:** If you find that you cannot access a specific command (or the help file for it), this may be because you don't have all the required packages and libraries installed. You can check which packages are installed in your copy of R using the `library()` command (**NOTE:** Even though you are checking for installed packages, you do this with a command with the term `library` in it). This will open a new window called R PACKAGES AVAILABLE containing a list of all the installed packages. If you are missing a package that you require to use a specific command, you will need to download and install it using the `install.packages` command. If you wish to check which libraries have been loaded into a specific project, you can use the `(.packages())` command (**NOTE:** Even though you are checking for installed libraries, you do this with a command with the term `packages` in it). This will return a list of all the libraries that have been loaded into your current R project. If the library containing the command you wish to use has not been loaded into R, you can use the `library` command to load it into your current R project.

5. **Look For A Solution And/Or Advice Online:** There will be occasions when you simply cannot work out the root cause of a problem or you just don't understand how to resolve an error message you are getting when you try to do something in R. In these situations, the best solution is usually to either use an internet search engine or post a question on one of the many online forums available for asking questions about using R and about how to do statistics in it, such as Stack Overflow (*www.stackoverflow.com*).

--- Chapter Three ---

Preparing Biological Data For Statistical Analysis Using R

The first step in any data analysis is to import your data into R, check them for errors, organise them by dividing them into subsets or by joining information from different sources together, and summarise them to help you understand what your data set contains. These basic actions not only allow you to get to know your data better, they can also help you to identify any issues there might be with them. This means the time you spend doing this will be more than repaid by the time saved trying to solve problems you may encounter later on if you do not. Thus, in this chapter, you will learn about the various ways you can import a data set into R, and then how you can check it for errors, divide a data set into subsets, join different data sets together and summarise the information a data set contains.

Before you start the exercises in this chapter, you first need to create a WORKING DIRECTORY folder on your computer and load the necessary data into it. To do this on a computer with a Windows operating system, open Windows Explorer and navigate to the location where you would like to create the folder (such as your C:\ drive or your DOCUMENTS folder). Next, right click anywhere in this location and select NEW> FOLDER. Now call this folder STATS_FOR_BIOLOGISTS_ONE by typing this into the folder name section to replace what it is currently called (which will most likely be NEW FOLDER). To create a WORKING DIRECTORY folder on a computer running a Mac operating system, open Finder and navigate to the location where you would like to create the folder (such as your DOCUMENTS folder or your DESKTOP). Next, click on FILE> NEW FOLDER, and then type the name STATS_FOR_BIOLOGISTS_ONE before pressing the ENTER key on your keyboard.

Once you have created your WORKING DIRECTORY folder, you are ready to download the data sets you will use for the exercises in this workbook from *www.gisinecology.com/stats-for-biologists-1*. After you have downloaded the compressed folder containing the required data by following the instructions provided on that page, you need to extract all the data files

from it and copy them into the folder called STATS_FOR_BIOLOGISTS_ONE that you have just created.

Next, you need to check that the required data have been extracted to the correct folder. If you are using a computer with a Windows operating system, you can use Windows Explorer to open your newly created WORKING DIRECTORY folder and examine its contents. If all the files from the compressed folder are present in it (there should be a total of 21 of them), you can click on the folder icon at the left hand end of the ADDRESS BAR at the top of the WINDOWS EXPLORER window to reveal its full address. Write this address down as you will need it to set this folder as your WORKING DIRECTORY during the exercises provided in this workbook (see pages 12 and 13 for details of how to modify folder addresses so they will be recognised by R).

If you are using a computer with a Mac operating system, you can use Finder to open your newly created WORKING DIRECTORY folder and examine its contents. If all the required data files are present in it (there should be a total of 21 of them), select this folder in Finder and then press the CMD and I keys on your keyboard at the same time. This will open the GET INFO window where you will find its address (which is also called the pathway). Write this address down somewhere as you will need it to set this folder as your WORKING DIRECTORY during the exercises provided in this workbook (see pages 12 and 13 for details of how to modify folder addresses so they will be recognised by R).

After you have loaded the required data into your WORKING DIRECTORY folder, you can open RGUI or RStudio, depending on which option you wish to use (see Chapter 2 for more details). Once you have opened your preferred R user interface, you need to create a file called CHAPTER_THREE_EXERCISES where you will save the results of your analyses from your R CONSOLE window as you work through this chapter. To do this using RGUI, click on the FILE menu and select SAVE WORKSPACE. To do this in RStudio, click on SESSION and select SAVE WORKSPACE AS. In both cases, save it as a WORKSPACE file with the name CHAPTER_THREE_EXERCISES.RDATA in your WORKING DIRECTORY folder (this will be the one called STATS_FOR_BIOLOGISTS_ONE that you have just created). If you are using RStudio, you will also want to save the contents of your SCRIPT EDITOR window (where you will enter and edit the R code you will use to carry out specific commands). To do this, click on the FILE menu and select SAVE AS. Save your file as an R SCRIPT file with the name CHAPTER_THREE_EXERCISES.R in your WORKING DIRECTORY folder. As you work through the exercises in this chapter, remember to regularly save the contents of your R CONSOLE window (which will contain the R objects you have created up to that point) to your WORKSPACE file and, if you are using RStudio, the contents of your SCRIPT EDITOR window to your R SCRIPT file.

Finally, you need to remove any data that are currently held in R's temporary memory. To do this, enter the following command into R (if you wish to copy and paste this command, the required code is directly below the text CODE BLOCK 1 in the document called R_CODE_BASIC_STATS_WORKBOOK.DOC that is included in the compressed folder you just downloaded):

```
rm(list=ls())
```

If you are using RGUI, you can simply type or paste this code after the command prompt at the bottom of the R CONSOLE window (it looks like this: >) and then press the ENTER key on your keyboard to run it. If you are using RStudio, you can type or paste this command into the SCRIPT EDITOR window (the upper left hand window). To run this command, select it and then click on the RUN button at the top of this window. This will run it in the R CONSOLE window (the lower left hand one in the main RStudio user interface). You are now ready to start the exercises in this chapter.

EXERCISE 1.1: HOW TO IMPORT DATA INTO R:

Data can be imported into R from a variety of different file formats. This includes comma separated value (.CSV) files, tab delimited (.TXT) files, data copied to your operating system's clipboard and spreadsheet files. In this exercise, you will learn how import data into R from each of these different file formats, starting with the comma separate value file format. A .CSV file is a file where the data held in different columns are separated by a comma (,) or, when a computer's operating system is set to use commas as the decimal separators, by a semicolon (;). It is one of the most common file formats used to transfer biological data from one program to another, including importing data from spreadsheet software (such as Microsoft Excel or OpenOffice Calc) into R.

The comma separated value file you will import into R in the first part of this exercise is called `blue_tit_occupancy_data.csv`. It contains data on the occupancy of nest boxes by blue tits, a small hole-nesting bird species, along with information on the location of each box and the land elevation at that location. To import the data from this .CSV file into R, work through the following flow diagram:

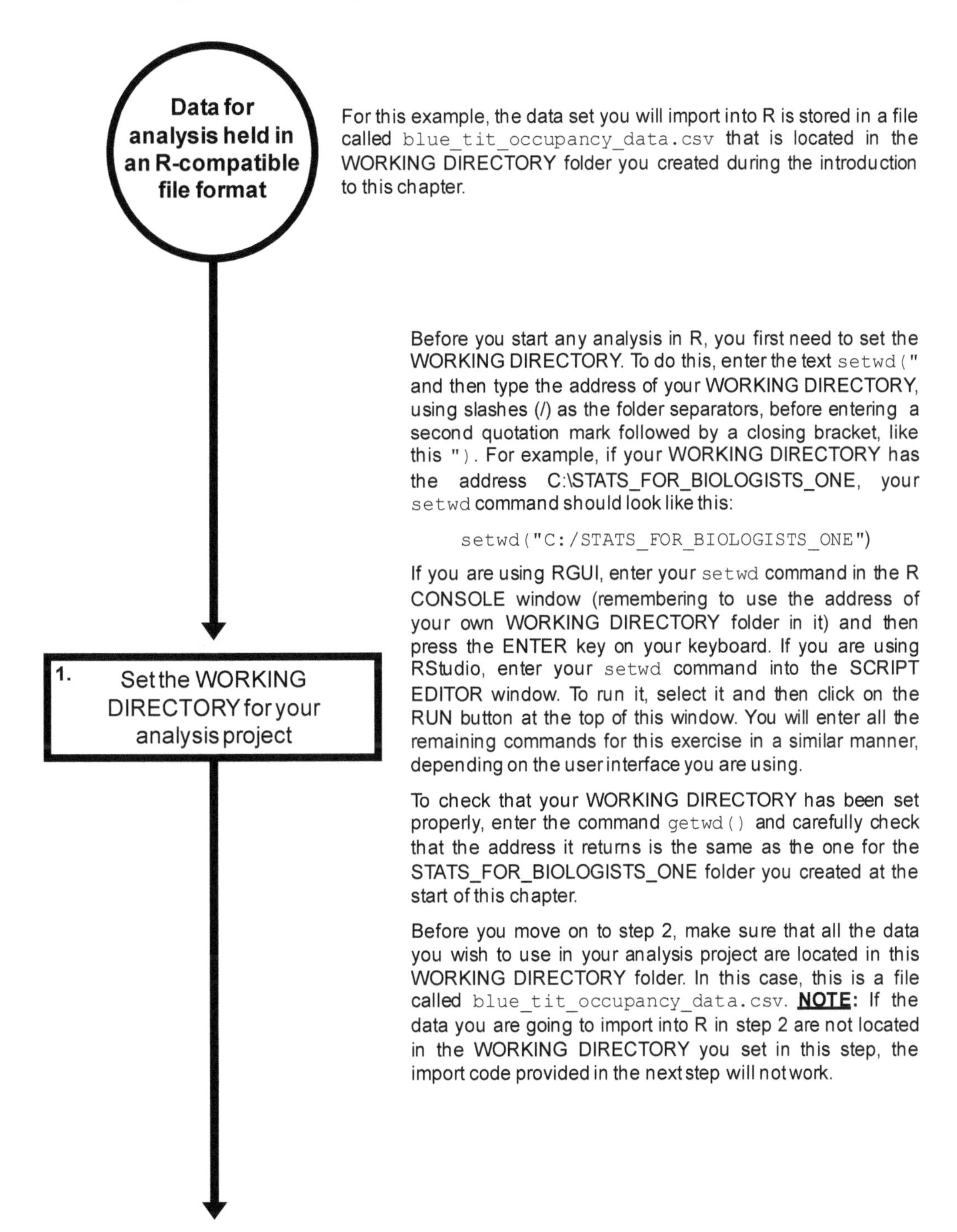
Data for analysis held in an R-compatible file format
For this example, the data set you will import into R is stored in a file called blue_tit_occupancy_data.csv that is located in the WORKING DIRECTORY folder you created during the introduction to this chapter.
1. Set the WORKING DIRECTORY for your analysis project
Before you start any analysis in R, you first need to set the WORKING DIRECTORY. To do this, enter the text setwd(" and then type the address of your WORKING DIRECTORY, using slashes (/) as the folder separators, before entering a second quotation mark followed by a closing bracket, like this "). For example, if your WORKING DIRECTORY has the address C:\STATS_FOR_BIOLOGISTS_ONE, your setwd command should look like this:
setwd("C:/STATS_FOR_BIOLOGISTS_ONE")
If you are using RGUI, enter your setwd command in the R CONSOLE window (remembering to use the address of your own WORKING DIRECTORY folder in it) and then press the ENTER key on your keyboard. If you are using RStudio, enter your setwd command into the SCRIPT EDITOR window. To run it, select it and then click on the RUN button at the top of this window. You will enter all the remaining commands for this exercise in a similar manner, depending on the user interface you are using.
To check that your WORKING DIRECTORY has been set properly, enter the command getwd() and carefully check that the address it returns is the same as the one for the STATS_FOR_BIOLOGISTS_ONE folder you created at the start of this chapter.
Before you move on to step 2, make sure that all the data you wish to use in your analysis project are located in this WORKING DIRECTORY folder. In this case, this is a file called blue_tit_occupancy_data.csv. NOTE: If the data you are going to import into R in step 2 are not located in the WORKING DIRECTORY you set in this step, the import code provided in the next step will not work.

2. Load your data into R using the `read.table` command

The `read.table` command provides the easiest way to load data held in a .CSV file into R so you can analyse it. To do this for the data being used in this example, enter the following command into R:

```
blue_tit_data <- read.table(file=
"blue_tit_occupancy_data.csv", sep=",",
as.is=FALSE, header=TRUE)
```

This code has to be entered exactly as it is written here or it will not work. If you wish to use the copy-and-paste approach for entering this command, copy the text directly below CODE BLOCK 2 in the document R_CODE_BASIC_STATS_WORKBOOK.DOC and paste it into R.

This command will create a new object in R called `blue_tit_data` which will contain the data from the specified .CSV file. To load a different .CSV file into R, all you need to do is change the file name in the `file` argument to the name of the one you wish to import. However, the specified file must be located in the WORKING DIRECTORY you set in step 1 of this flow diagram. You can also use whatever name you wish for the R object that will be created by this command. To do this, simply replace `blue_tit_data` at the start of the above command with the name you wish to use for it. **NOTE:** If your .CSV data set uses a semi-colon as the decimal separator, you would need to replace the `sep=","` argument with `sep=";"`.

3. Check the data have loaded into R correctly by checking the names of its columns, the number of rows it contains and by viewing it

Whenever you import a data set into R, you need to check that it has been loaded correctly. First, you need to check that all the required columns are present in the R object you just created. To do this, enter the following command into R:

```
names(blue_tit_data)
```

This is CODE BLOCK 3 in the document R_CODE_BASIC_STATS_WORKBOOK.DOC. This command will return the names used for each column in the R object called `blue_tit_data` created in step 2. For this example, the names should be: `box_number`, `latitude`, `longitude`, `occupied`, `elevation` and `el_cat`.

Next, you should check the number of rows in your R object to make sure that the entire data set has been successfully imported into R. To do this, you need to specify one of the columns within your newly created R object. For this example, you will use the column called `box_number` in the R object called `blue_tit_data`. To count the number of rows in this column in this R object, enter the following command into R:

```
length(blue_tit_data$box_number)
```

This is CODE BLOCK 4 in the document R_CODE_BASIC_STATS_WORKBOOK.DOC. For the data set being used in this example, the number of rows this command returns should be 198.

Finally, you should view the contents of your newly created R object (called `blue_tit_data` in this example) using the `View` command (**NOTE:** Unlike most commands in R, this command begins with a capital letter). This is done by entering the following code into R:

```
View(blue_tit_data)
```

This is CODE BLOCK 5 in the document R_CODE_BASIC_STATS_WORKBOOK.DOC. This command will open a DATA VIEWER window where you can examine your data set and check that the correct data have been loaded into R.

At the end of the first part of this exercise, the last few lines of your R CONSOLE window should look like this (**NOTE:** Your WORKING DRECTORY folder will have a different address to the one shown here if it has been created in a different location on your computer):

```
> setwd("C:/STATS_FOR_BIOLOGISTS_ONE")
> getwd()
[1] "C:/STATS_FOR_BIOLOGISTS_ONE"
> blue_tit_data <- read.table(file="blue_tit_occupancy_data.csv", sep=",", header=TRUE)
> names(blue_tit_data)
[1] "box_number" "latitude"   "longitude"  "occupied"   "elevation"
[6] "el_cat"
> length(blue_tit_data$box_number)
[1] 198
> View(blue_tit_data)
> |
```

While the contents of the DATA VIEWER window should look like this:

	box_number	latitude	longitude	occupied	elevation	el_cat
1	138	56.12644	-4.617821	0	23.27828	20 to 30
2	139	56.12641	-4.618161	0	21.50202	20 to 30
3	141	56.12622	-4.617965	0	20.90909	20 to 30
4	144	56.12598	-4.616986	0	10.00000	0 to 10
5	143	56.12607	-4.616885	0	15.00000	10 to 20
6	142	56.12622	-4.616998	1	20.00000	10 to 20
7	137	56.12642	-4.617008	0	21.58028	20 to 30
8	33	56.13024	-4.614165	0	30.82850	30 to 40
9	30	56.13059	-4.614695	0	31.48324	30 to 40
10	24	56.13129	-4.615251	1	22.87749	20 to 30
11	48	56.12906	-4.614811	1	25.00000	20 to 30
12	13	56.13031	-4.616208	0	42.50000	40 to 50
13	14	56.13052	-4.616450	0	40.00000	30 to 40
14	19	56.13106	-4.616647	0	32.61233	30 to 40
15	301	56.13181	-4.617249	1	32.92912	30 to 40
16	44	56.13033	-4.617238	1	60.00000	50 or more
17	4	56.12952	-4.615516	0	40.00000	30 to 40
18	3	56.12927	-4.615110	1	32.50000	30 to 40
19	300	56.13214	-4.616719	1	25.14756	20 to 30
20	302	56.13208	-4.617021	0	28.33333	20 to 30

While the `read.table` command works for most .CSV files, there may be some occasions where it does not work with a particular one. In such instances, you can modify the code used in step 2 of the above flow diagram to use the more specific `read.csv` command rather than the more generic `read.table` command. The modified version of this command would look like the one provided at the top of the next page (required modifications highlighted in **bold**).

```
blue_tit_data <- read.csv(file="blue_tit_occupancy_
    data.csv", sep=",", as.is=FALSE, header=TRUE)
```

The `read.table` command can be modified in a variety of different ways to allow you to import data sets saved in other file formats into R. When you do this, the main argument that you need to change is the `sep` argument (this tells R what symbol the file uses to separate the data in different columns of the same row or line). To explore how you can do this, in the next part of this exercise you will modify the `read.table` command from step 2 in the above flow diagram to import the contents of a tab delimited file called `blue_tit_occupancy_data.txt` into R (this file contains exactly the same data as the .CSV file you previously imported). To do this, you should modify the code so that it looks like this (**NOTE:** Make sure that you use a backslash (\) rather than the more usual slash (/) in the `sep` argument):

```
blue_tit_data <- read.table(file="blue_tit_occupancy_
    data.txt", sep="\t", as.is=FALSE, header=TRUE)
```

You can modify this code either by editing it in the R CONSOLE window of RGUI or through the SCRIPT EDITOR window of RStudio (depending on which interface you are using). If you are entering commands directly into the R CONSOLE window, you can use the UP arrow on your keyboard to bring commands you have previously run during the same session back on to the command line of this window, and then use the LEFT and RIGHT arrows to scroll through and edit them. In this case, use the UP arrow to bring the previous version of the `read.table` command back onto the command line and edit it so that it looks like the one above. Once you have finished modifying your `read.table` command, you can run it by pressing the ENTER key on your keyboard. If you are using RStudio, you can copy and paste the original `read.table` command in the SCRIPT EDITOR window before editing the new version to include the required modifications. Once you have done this, select the modified version of the command and click on the RUN button to run it in the R CONSOLE window.

Once you have run this new version of the `read.table` command, you need to run the `names`, `length` and `View` commands from step 3 of the above flow diagram again to check that the data have been loaded correctly. The results of these commands and the imported data should look exactly the same as when you used the original `read.table` command to import the contents of the above .CSV file.

Similarly, if you wish to import a data set that has been copied to the clipboard of your operating system, you can use the same command format used for importing tab delimited data. However, you will need to change the file name in the command to allow it to read in the contents of your operating system's clipboard rather reading the contents of a specific file. The modified version of the `read.table` command that would allow you to import data from your clipboard (rather than from a tab delimited file) would look like this:

```
blue_tit_data <- read.table(file="clipboard", sep="\t",
                 as.is=FALSE, header=TRUE)
```

NOTE: If you are using a computer with a Mac operating system, you would need to modify the clipboard address in the `file` argument so that it reads `file=` **`pipe("pbpaste")`** rather than `file="`**`clipboard`**`"`.

To practice using this method of importing data into R, open the file called `blue_tit_occupancy_data.xls` with a spreadsheet programme and copy its contents to the clipboard of your operating system (this file contains exactly the same data as the .CSV file with the same name). Once you have done this, you can run the above version of the `read.table` command to import it into your current project, followed by the `names`, `length` and `View` commands from step 3 of the above flow diagram to check that it has been imported correctly into R. The results of these commands and imported data should look exactly the same as before.

More information about how you can modify the `read.table` command, including additional arguments that you can use with it to refine exactly which data are imported from the original file, can be found at *www.rdocumentation.org/packages/utils/versions/3.6.1/topics/read.table*.

If you wish to import data from a spreadsheet file, you cannot use the `read.table` command. Instead, you need to use an alternative tool, such as one of those in the ReadXL package. However, before you can use these tools, you will first need to install this package into your copy of R. This is done using the following command (this is CODE BLOCK 6 in the document R_CODE_BASIC_STATS_WORKBOOK.DOC):

```
install.packages("readxl")
```

To install the ReadXL package, enter this code into R and then follow any instructions which appear. Once this package has been successfully downloaded and installed, you can use its `read_excel` command to import data from an Excel spreadsheet into R. To practice using this tool, you can use it to import the contents of the file `blue_tit_occupancy_data.xls` into R. To do this, you will need to modify the code in step 2 of the above flow diagram to reflect that it should use a different data import tool and file, including loading the library for the package that contains the tool you wish to use. This is done by adding the `library` command at the start of this block of code. Once you have modified the code for step 2, it should look like this:

```
library(readxl)
blue_tit_data_xl <- read_excel("blue_tit_occupancy_
data.xls", sheet=1)
```

After you have run this modified version of the original code for this step, repeat the `names`, `length` and `View` commands from step 3 of the above flow diagram (remembering to change the R object name in them to `blue_tit_data_xl`) to check that the data have been loaded correctly. The imported data should look exactly the same as the data imported from the .CSV file.

More information about how you can use the tools in the ReadXL package to import spreadsheet data into R, including additional arguments that you can use with it to refine exactly which data are imported from the original file, can be found at *www.rdocumentation.org/packages/readxl/versions/1.3.*

EXERCISE 1.2: HOW TO CHECK A DATA SET FOR ERRORS IN R:

Whenever you import a data set into R, and before you start analysing it, it is important that you check it for errors. Potential errors can include missing rows of data (if the data have not been prepared or imported properly), missing variables (again, if the data have not been prepared or imported properly), missing values, data stored in the wrong data format, and data which have not been assigned the correct codes for categorical variables. Such errors can cause major issues with your data analysis and, if they are present, they need to be corrected before you proceed. The data that you will use for the first part of this exercise is the R object called `blue_tit_data` created in Exercise 1.1, and the instructions for this exercise assume you have successfully completed all the steps in the data import workflow. This means you will need to have completed that exercise before you can do this one. **NOTE:** The same error-checking procedure applies regardless of how you have imported your data into R. To learn how to error check data you have previously imported into R, work through the following flow diagram:

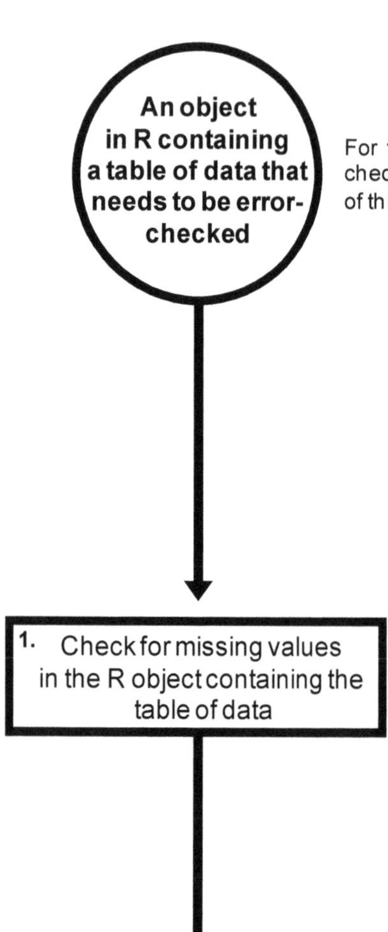
An object in R containing a table of data that needs to be error-checked
1. Check for missing values in the R object containing the table of data

For this example, the R object containing the data you will error-check is called `blue_tit_data` and it was created in Exercise 1.1 of this workbook.

When error-checking a data set, the first thing you should do is to look for missing values. To do this for the R object being error-checked in this example, enter the following commands into R:

```
sum(is.na(blue_tit_data))
is.na(blue_tit_data)
```

This code has to be entered exactly as it is written here or it will not work. If you wish to use the copy-and-paste approach for entering these commands, copy the text directly below CODE BLOCK 7 in the document R_CODE_BASIC_STATS_WORKBOOK.DOC and paste it into R. The first of these two commands will calculate the total number of missing values in your data set, while the second will return a table of your entire data set that will allow you to identify exactly which values are missing (these will be marked with the word `TRUE`). **NOTE:** If, when you run the `is.na` command, you get a warning message starting with the words `reached getOption("max.print")`, enter the command `options(max.print=999999)` and then re-run the `is.na` command. For the R object being examined in this example (called `blue_tit_data`), you should find that there are no missing values.

2. Check the structure of the data within the R object containing the table of data

To check the structure of the data within an R object, you can use the str command. This will return a table with a list of the variables, their type (i.e. whether they contain whole numbers, numbers with decimal places or categories) and examples of the first few values within it. To display this information for the R object called blue_tit_data, enter the following command into R:

```
str(blue_tit_data)
```

This is CODE BLOCK 8 in the document R_CODE_BASIC_STATS_WORKBOOK.DOC. For the R object being examined in this example, the table should contain a list of six variables, and it should correctly identify which contain whole numbers (int), which contain numbers with decimal places (num) and which contain categories (Factor).

3. Check which categories are included for categorical variables

While the str command will tell you which columns contain categorical variables (also known as factors), it doesn't necessarily provide a full list of the categories each one contains. For this example, you will check the full list of categories included in the categorical variable held in the column called el_cat. To do this, enter the following command into R:

```
unique(blue_tit_data$el_cat)
```

This is CODE BLOCK 9 in the document R_CODE_BASIC_STATS_WORKBOOK.DOC. For the R object being used for this example, this should return the six categories present in the el_cat (short for elevation categories) column. These are: 0 to 10, 10 to 20, 20 to 30, 30 to 40, 40 to 50 and 50 or more.

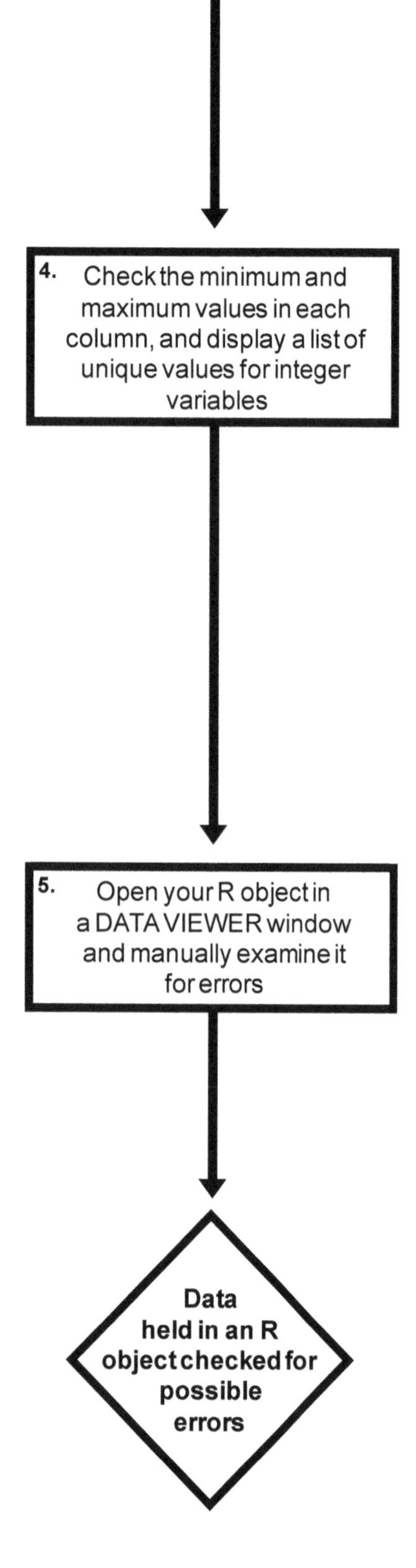

For numerical variables, such as those containing whole numbers (also known as integers), it is useful to be able to check the minimum and maximum values as well as returning a list of all the unique values contained in them. For this example, you will do these checks for the numerical variable held in the column called `occupied`. This column should contain a `0` for all rows of data that represent nest boxes not occupied by blue tits, and a `1` for all the rows of data that represent nest boxes occupied by this species. If it contains any other values, these will be errors and would cause problems with any data analysis that assumes this column contains only 1s and 0s (known as binomial data). To do this, enter the following commands into R:

```
min(blue_tit_data$occupied)
max(blue_tit_data$occupied)
unique(blue_tit_data$occupied)
```

This is CODE BLOCK 10 in the document R_CODE_BASIC_STATS_WORKBOOK.DOC. For the R object being used for this exercise, this should return a minimum value of `0` (the first number returned), a maximum value of `1` (the second value returned) and a list containing only the values `0` and `1` (the last line returned).

The last step in error-checking a data set in R is to manually scroll through it in a DATA VIEWER window and check it for errors. To do this for the R object being examined in this example, enter the following command into R (**NOTE:** Unlike most R commands, this command begins with a capital letter):

```
View(blue_tit_data)
```

This is CODE BLOCK 11 in the document R_CODE_BASIC_STATS_WORKBOOK.DOC. This will open a DATA VIEWER window and display the contents of the specified R object within it (in this case, the object called `blue_tit_data` created in Exercise 1.1)

At the end of the first part of this exercise, the last few lines of your R CONSOLE window should look similar to this (**NOTE:** The exact content of this window may vary slightly depending on whether you are using RStudio or RGUI, and on your version of R):

```
[193,]      FALSE    FALSE      FALSE    FALSE      FALSE  FALSE
[194,]      FALSE    FALSE      FALSE    FALSE      FALSE  FALSE
[195,]      FALSE    FALSE      FALSE    FALSE      FALSE  FALSE
[196,]      FALSE    FALSE      FALSE    FALSE      FALSE  FALSE
[197,]      FALSE    FALSE      FALSE    FALSE      FALSE  FALSE
[198,]      FALSE    FALSE      FALSE    FALSE      FALSE  FALSE
> str(blue_tit_data)
'data.frame':   198 obs. of  6 variables:
 $ box_number: int  138 139 141 144 143 142 137 33 30 24 ...
 $ latitude  : num  56.1 56.1 56.1 56.1 56.1 ...
 $ longitude : num  -4.62 -4.62 -4.62 -4.62 -4.62 ...
 $ occupied  : int  0 0 0 0 0 1 0 0 0 1 ...
 $ elevation : num  23.3 21.5 20.9 10 15 ...
 $ el_cat    : Factor w/ 6 levels "0 to 10","10 to 20",..: 3 3 3 1 2 2 3 4 4 3 ...
> unique(blue_tit_data$el_cat)
[1] 20 to 30   0 to 10    10 to 20   30 to 40   40 to 50   50 or more
Levels: 0 to 10 10 to 20 20 to 30 30 to 40 40 to 50 50 or more
> min(blue_tit_data$occupied)
[1] 0
> max(blue_tit_data$occupied)
[1] 1
> unique(blue_tit_data$occupied)
[1] 0 1
> View(blue_tit_data)
> |
```

While the contents of the DATA VIEWER window should look like this:

	box_number	latitude	longitude	occupied	elevation	el_cat
1	138	56.12644	-4.617821	0	23.27828	20 to 30
2	139	56.12641	-4.618161	0	21.50202	20 to 30
3	141	56.12622	-4.617965	0	20.90909	20 to 30
4	144	56.12598	-4.616986	0	10.00000	0 to 10
5	143	56.12607	-4.616885	0	15.00000	10 to 20
6	142	56.12622	-4.616998	1	20.00000	10 to 20
7	137	56.12642	-4.617008	0	21.58028	20 to 30
8	33	56.13024	-4.614165	0	30.82850	30 to 40
9	30	56.13059	-4.614695	0	31.48324	30 to 40
10	24	56.13129	-4.615251	1	22.87749	20 to 30
11	48	56.12906	-4.614811	1	25.00000	20 to 30
12	13	56.13031	-4.616208	0	42.50000	40 to 50
13	14	56.13052	-4.616450	0	40.00000	30 to 40
14	19	56.13106	-4.616647	0	32.61233	30 to 40
15	301	56.13181	-4.617249	1	32.92912	30 to 40
16	44	56.13033	-4.617238	1	60.00000	50 or more
17	4	56.12952	-4.615516	0	40.00000	30 to 40
18	3	56.12927	-4.615110	1	32.50000	30 to 40
19	300	56.13214	-4.616719	1	25.14756	20 to 30
20	302	56.13208	-4.617021	0	28.33333	20 to 30

One of the few limitations of R is that it doesn't come with efficient data editing tools. As a result, if you find errors in your data set, by far the easiest way for less experienced R users to correct them is to edit the contents of the original file using the software it was created in, such as Microsoft Excel or OpenOffice Calc, or using a text editor program. After you have done this, you can re-import the corrected data set into R using the appropriate import command (see Exercise 1.1). To explore how you can do this in the next part of this exercise, you first need to import a new data set called `error_checking_data.csv` by entering the following command into R:

```
blue_tit_data <- read.table(file="error_checking_
  data.csv", sep=",", as.is=FALSE, header=TRUE)
```

The new data set imported by this command is a second version of the data set imported in Exercise 1.1 and that was error-checked when you worked through the above flow diagram. However, some errors have purposely been added to it which you will need to detect and correct. Once this new data set has been imported into R, you can repeat steps 1 to 5 of the above flow diagram (using the same R commands) to error-check it and compare the results you got from the two error-checks you have run. When you do this, you will see that the maximum value for the column called `occupied` in the second version of the data set is `11` (rather than `1`, as if should be) and the list of unique values for this column contains the values `0`, `1` and `11`. Since this column should only contain the values `0` and `1` (as detailed in step 4 of the above flow diagram), there is clearly a problem here. Specifically the problem is that in the file `error_checking_data.csv`, a value of `11` has been entered into the `occupied` column for several rows, instead of a value of `1` (this is a type of error that is very easy to make during data entry, but they can have a negative impact on your analysis if they are not spotted and corrected).

To correct these errors, open the original .CSV file (called `error_checking_data.csv`) in a spreadsheet program, such as Microsoft Excel. Once this file is open, find all the rows of data with a value of `11` in the column called `occupied` and change them to a value of `1`. After you have corrected these errors, save the changes you have made to this .CSV file and re-import it into R by re-running the `read.table` command provided at the top of this page.

Finally, you can run steps 1 to 5 of the above flow diagram once again to ensure that all the errors in the data set have been detected and corrected. When you examine the results you obtain from re-running these checks on the edited data set, you should find that the maximum value for the column called `occupied` is now `1` (as it should be, and not `11` as it was) and the list of unique values for this column now only contains the values `0` and `1`, again as it should. Thus, you have now confirmed that the errors originally present in the data set `error_checking_data.csv` have been corrected. If you found any further errors at this point, you would keep repeating this editing and error-checking procedure until you are confident that all the errors in your data set have been dealt with appropriately.

While the workflow outlined above is primarily aimed at editing a data set to remove any errors that you have identified during error-checking, the same approach can be used for any other type of editing you need to carry out. This includes renaming columns, adding new columns, grouping continuous variables into categories or reclassifying data. While all of these processes can be done in R, it is usually quicker and easier for novice R users to do them in other types of software, such as a spreadsheet program.

EXERCISE 1.3: HOW TO DIVIDE A DATA SET INTO SUBSETS IN R:

There are many occasions when you will wish to divide your data into subsets before you analyse them. For example, you might wish to create a subset containing only the data for one species in a multi-species data set, or create a subset containing data from a single month or year from a larger data set. Dividing data into subsets is relatively easy in R as long as your data include a column that contains a variable which can be used to identify the subset you wish to extract. While you can do it in R, if your data does not already contain such a column, the easiest way to add one is to use a spreadsheet program to create it before you bring your data into R. The data set that you will extract a subset from in this exercise is contained in the R object called `blue_tit_data` originally created in Exercise 1.1. To do this, work through the flow diagram that starts at the top of the next page.

An object in R containing a table of data from which you wish to extract a subset

For this example, the R object containing the data from which you will extract a subset is called `blue_tit_data` and it was created in Exercise 1.1 of this workbook.

1. Identify the name of the column which you wish to use to create your subset, and the values that are present in it

Before you can extract a subset of data from your existing data set, you need to check on the names used for the columns in it so you can identify the one you wish to use to create your subset. To do this, enter the following command into R:

```
names(blue_tit_data)
```

This code has to be entered exactly as it is written here or it will not work. If you wish to use the copy-and-paste approach for entering this command, copy the text directly below CODE BLOCK 12 in the document R_CODE_BASIC_STATS_WORKBOOK.DOC and paste it into R. For the data set being used in this example, this will return the names `box_number`, `latitude`, `longitude`, `occupied`, `elevation` and `el_cat`.

Once you have identified the name of the column which you wish to use to create your subset, you need to find out what values it contains. For this example, you will use the column `occupied` in the R object `blue_tit_data`. To find out what values are contained in this column, enter the following command into R:

```
unique(blue_tit_data$occupied)
```

This is CODE BLOCK 13 in the document R_CODE_BASIC_STATS_WORKBOOK.DOC. For the data set being used in this exercise, this will return two values: 1 (which indicates that a specific nest box was occupied by blue tits) and 0 (which indicates it was not occupied by blue tits). **NOTE:** If you get more than two values for this column, or if either of the values returned are not 0 or 1, you will need to return to Exercise 1.1 and re-import the original data set being used in this example.

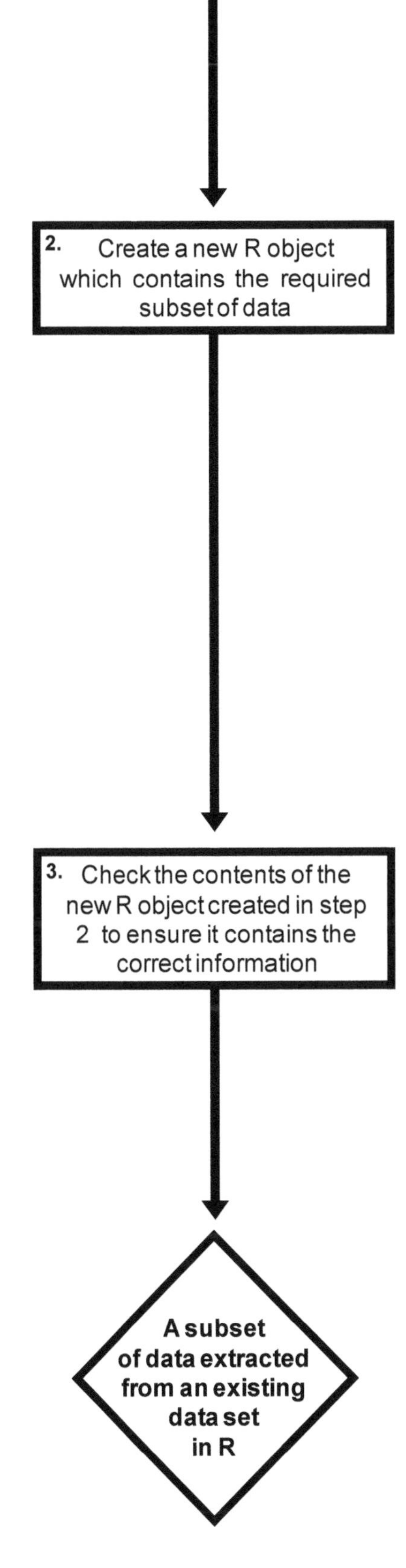

2. Create a new R object which contains the required subset of data
3. Check the contents of the new R object created in step 2 to ensure it contains the correct information
A subset of data extracted from an existing data set in R

To create a new object in R which contains the required subset of data, you need to use the `subset` command followed by the name of the original data set, the name of the column which contains the variable you wish to use to create the subset, and the values that you wish to extract from it. You will also need to specify the name you wish to use for the new R object this command will create. In this example, you will extract all the rows of data for the nest boxes occupied by blue tits from the data set `blue_tit_data`. These nest boxes can be identified by a value of 1 in the column called `occupied`. To do this, enter the following command into R:

```
occupied_boxes <- subset(blue_tit_data,
        occupied=="1")
```

This is CODE BLOCK 14 in the document R_CODE_BASIC_STATS_WORKBOOK.DOC. Once this command has finished running, it will appear as if nothing has happened. This is okay. You will check that it has worked properly in the next step.

To check the contents of your new R object, you can run some of the same error-checking procedures outlined in Exercise 1.2. These will allow you to examine your new data set and check that it only contains the required subset of data. To run these checks on the R object you have just created (called `occupied_boxes`), enter the following commands into R:

```
names(occupied_boxes)
sum(is.na(occupied_boxes))
is.na(occupied_boxes)
str(occupied_boxes)
unique(occupied_boxes$occupied)
View(occupied_boxes)
```

This is CODE BLOCK 15 in the document R_CODE_BASIC_STATS_WORKBOOK.DOC. Once these commands have finished running, you can examine the results. If you look in the R CONSOLE window, you will see that the column called `occupied` for the R object `occupied_boxes` only contains a value of 1, indicating that these data have been extracted correctly. Similarly, if you examine the DATA VIEWER window, you will see that all the data in the `occupied` column have a value of `1`, indicating that the R object created in step 2 only contains this subset of the original data.

At the end of the first part of this exercise, the last few lines of your R CONSOLE window should look like this:

```
166      FALSE    FALSE     FALSE    FALSE     FALSE  FALSE
167      FALSE    FALSE     FALSE    FALSE     FALSE  FALSE
171      FALSE    FALSE     FALSE    FALSE     FALSE  FALSE
174      FALSE    FALSE     FALSE    FALSE     FALSE  FALSE
176      FALSE    FALSE     FALSE    FALSE     FALSE  FALSE
180      FALSE    FALSE     FALSE    FALSE     FALSE  FALSE
181      FALSE    FALSE     FALSE    FALSE     FALSE  FALSE
184      FALSE    FALSE     FALSE    FALSE     FALSE  FALSE
186      FALSE    FALSE     FALSE    FALSE     FALSE  FALSE
190      FALSE    FALSE     FALSE    FALSE     FALSE  FALSE
192      FALSE    FALSE     FALSE    FALSE     FALSE  FALSE
193      FALSE    FALSE     FALSE    FALSE     FALSE  FALSE
196      FALSE    FALSE     FALSE    FALSE     FALSE  FALSE
> str(occupied_boxes)
'data.frame':   66 obs. of  6 variables:
 $ box_number: int  142 24 48 301 44 3 300 308 309 315 ...
 $ latitude  : num  56.1 56.1 56.1 56.1 56.1 ...
 $ longitude : num  -4.62 -4.62 -4.61 -4.62 -4.62 ...
 $ occupied  : int  1 1 1 1 1 1 1 1 1 1 ...
 $ elevation : num  20 22.9 25 32.9 60 ...
 $ el_cat    : Factor w/ 6 levels "0 to 10","10 to 20",..: 2 3 3 4 6 4 3 3 3 3 ...
> unique(occupied_boxes$occupied)
[1] 1
> View(occupied_boxes)
> |
```

While the contents of the DATA VIEWER window should look like this:

	row.names	box_number	latitude	longitude	occupied	elevation	el_cat
1	6	142	56.12622	-4.616998	1	20.00000	10 to 20
2	10	24	56.13129	-4.615251	1	22.87749	20 to 30
3	11	48	56.12906	-4.614811	1	25.00000	20 to 30
4	15	301	56.13181	-4.617249	1	32.92912	30 to 40
5	16	44	56.13033	-4.617238	1	60.00000	50 or more
6	18	3	56.12927	-4.615110	1	32.50000	30 to 40
7	19	300	56.13214	-4.616719	1	25.14756	20 to 30
8	21	308	56.13260	-4.617383	1	25.85823	20 to 30
9	22	309	56.13341	-4.617806	1	22.50000	20 to 30
10	23	315	56.13388	-4.618419	1	25.66380	20 to 30
11	27	304	56.13184	-4.618219	1	41.11111	40 to 50
12	30	307	56.13252	-4.617935	1	33.33333	30 to 40
13	33	73	56.12795	-4.616873	1	46.66667	40 to 50
14	36	94	56.12737	-4.617562	1	40.00000	30 to 40
15	42	108	56.12726	-4.619149	1	43.33333	40 to 50
16	48	99	56.12689	-4.618082	1	26.66667	20 to 30
17	53	112	56.12680	-4.619603	1	25.00000	20 to 30
18	54	114	56.12685	-4.620261	1	30.00000	20 to 30
19	55	118	56.12689	-4.620504	1	30.00000	20 to 30
20	58	191	56.12657	-4.619749	1	20.00000	10 to 20

As well as selecting data based on a specific value, you can also use the `subset` command to select data that are not equal to a specific value, that are more or less than a specific value or that fulfil multiple criteria. These different methods of selecting data can be implemented by modifying the selection argument of the `subset` command in step 2 of the above flow diagram. Originally, the selection argument included in this command was `occupied=="1"`, which meant that the data would be subsetted based on whether or not the records had a value of `1` for the column called `occupied` (**NOTE:** With the selection argument, categories for categorical variables are enclosed by quotation marks, while values for continuous variables are not). To select data that are not equal to a specific value for a categorical variable you would use `!=` in the selection argument, so if you used `el_cat!="0 to 10"` in the above example (instead of `occupied=="1"`), it would select all the nest boxes where the category in the `el_cat` column is not `0 to 10`. To select data that are greater or equal to a specific value for a continuous variable, you would use >=. Thus, if you use `elevation>=15` in the above example it would select all the nest boxes where the value in the `elevation` column is greater or equal to 15. To select data that are just greater than a specific value, you would use >, while to select data that are less than or equal to a specific value, you would use <=, and to select data that are less than a specific value, you would use <.

To select data based on more that one criteria, you would use an AND or an OR function. In the `subset` command, an AND function is set by using the ampersand symbol (`&`), so including the selection argument `el_cat=="0 to 10" & occupied=="1"` in the above command would select all nest boxes that have a value for the `el_cat` column of `0 to 10` and a value of `1` for the `occupied` column. An OR function is set using the `|` symbol, so including the selection argument `el_cat=="0 to 10"|el_cat=="10 to 20"` in the above command would select all the nest boxes that have a value for the `el_cat` column of `0 to 10` or `10 to 20`.

If, once you have subsetted your data, you wish to export it from R for use in a different software package, you can do this using the `write.table` command. For example, to export the R object called `occupied_boxes` created using the `subset` command in the above flow diagram as a .CSV file to your project's WORKING DIRECTORY folder, your `write.table` command would need to look like this:

```
write.table(occupied_boxes, file="occupied_
boxes.csv", sep=",")
```

EXERCISE 1.4: HOW TO JOIN DATA SETS TOGETHER IN R:

Data sets can be joined together in R if they have shared values in a common column. This is done using the `merge` command. Within this command, you can specify which data sets should be joined together, what columns should be used to join them and how this join should be carried out. As an example of how to do this, in this exercise you will join two data sets together. The first of these data sets is contained in the R object `blue_tit_data` first created in Exercise 1.1, while the second is a new data set called `nestbox_breeding_data.csv`. This second data set contains information about the breeding success of all the species (and not just blue tits) recorded from the same set of nest boxes. These data sets will be joined together based on the values in a column called `box_number` that contains a unique identification number for each individual nest box that is shared by both data sets. To do this join, work through the following flow diagram:

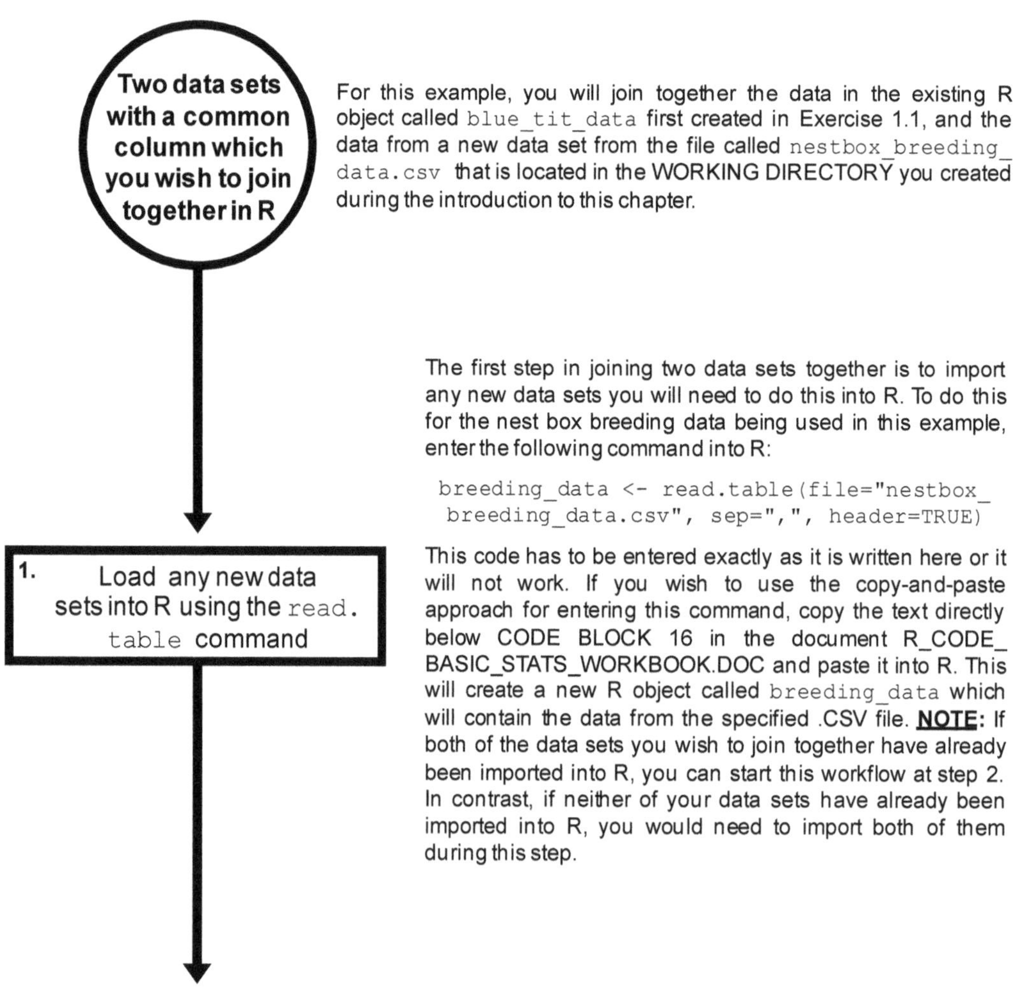

Two data sets with a common column which you wish to join together in R
For this example, you will join together the data in the existing R object called blue_tit_data first created in Exercise 1.1, and the data from a new data set from the file called nestbox_breeding_data.csv that is located in the WORKING DIRECTORY you created during the introduction to this chapter.
The first step in joining two data sets together is to import any new data sets you will need to do this into R. To do this for the nest box breeding data being used in this example, enter the following command into R:
breeding_data <- read.table(file="nestbox_breeding_data.csv", sep=",", header=TRUE)
1. Load any new data sets into R using the read.table command
This code has to be entered exactly as it is written here or it will not work. If you wish to use the copy-and-paste approach for entering this command, copy the text directly below CODE BLOCK 16 in the document R_CODE_BASIC_STATS_WORKBOOK.DOC and paste it into R. This will create a new R object called breeding_data which will contain the data from the specified .CSV file. NOTE: If both of the data sets you wish to join together have already been imported into R, you can start this workflow at step 2. In contrast, if neither of your data sets have already been imported into R, you would need to import both of them during this step.

2. Check the names of the two data sets to help you identify the common column which you can use to join the data sets together.

To check the column names of the two data sets you will join together in this example, enter the following commands into R:

```
names(blue_tit_data)
names(breeding_data)
```

This is CODE BLOCK 17 in the document R_CODE_BASIC_STATS_WORKBOOK.DOC. For the data set in the R object called `blue_tit_data`, the column names returned should be `box_number`, `latitude`, `longitude`, `occupied`, `elevation` and `el_cat`. For the data set in the R object called `breeding_data`, the column names returned should be `box_number`, `species`, `clutch_size`, `no_chicks_hatched` and `no_chicks_fledged`.

From these two lists of columns, you can identify any common column names between the two data sets which could be used to join them together. For the data sets being used in this example, the common column that will be used for the join is called `box_number`, which contains a unique identification number for each nest box that is shared by both data sets.

3. Join the two data sets together based on the values in the common column.

In R, data sets can be joined together based on values in a common column using the `merge` command. To join the two data sets being used in this example based on the common column called `box_number`, enter the following command into R:

```
combined_dataset <- merge(x=blue_tit_data,
    y=breeding_data, by="box_number",
        all.x=TRUE).
```

This is CODE BLOCK 18 in the document R_CODE_BASIC_STATS_WORKBOOK.DOC. This command will create an R object called `combined_dataset`. This will contain all the information from the data set called `breeding_data` that has a value in the column called `box_number` which is also found in the same column in data set called `blue_tit_data`. **NOTE**: If you wanted to include all the data from both data sets, regardless of whether the `box_number` value was found in the first data set, you would replace the term `all.x=TRUE` with `all=TRUE`. Similarly, if you wished to only include data with values for the common column found in both data sets, you would use the term `all=FALSE`.

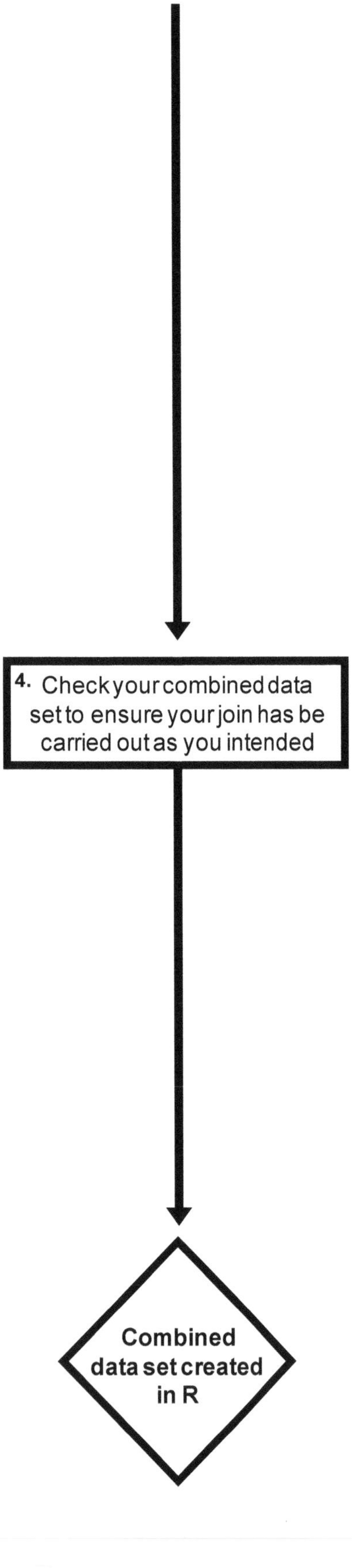

4. Check your combined data set to ensure your join has be carried out as you intended

Combined data set created in R

To ensure that your join has been carried out as you intended, you need to carry out three checks. Firstly, you need to look at the column names present in both your original data sets and your new combined data set. To do this for the data sets being used in this example, enter the following commands into R:

```
names(blue_tit_data)
names(breeding_data)
names(combined_dataset)
```

This is CODE BLOCK 19 in the document R_CODE_BASIC_STATS_WORKBOOK.DOC. This set of commands will return the names of the columns in all three data sets. Check that all the column names included in the data set called `blue_tit_data` (the first list of names returned) and the data set called `breeding_data` (the second list of names returned) are also present in the data set called `combined_dataset` created by the `merge` function (the final list of names returned).

Secondly, you should look at the number of records in both your original data sets and your new combined data set. To do this for the data sets being used in this example, enter the following code into R:

```
length(blue_tit_data$box_number)
length(breeding_data$box_number)
length(combined_dataset$box_number)
```

This is CODE BLOCK 20 in the document R_CODE_BASIC_STATS_WORKBOOK.DOC. This set of commands will return the number of rows in `blue_tit_data` (the first number returned), `breeding_data` (the second number returned) and the `combined_dataset` object created by the `merge` function (the final number returned). Since you specified that you wish to keep all records from the first data set (by using the term `all.x=TRUE` in the `merge` function), the number of rows in `blue_tit_data` and the `combined_dataset` object should be the same. If you had used the term `all.y=TRUE`, then the number of rows of `breeding_data` and `combined_dataset` should be the same, while if you used the terms `all=TRUE` or `all=FASLE`, you may get a different number of rows for all three data sets.

Finally, you need to view your combined data set, so you can manually check that it contains the data it should. For this example, this is done by entering the following command into R:

```
View(combined_dataset)
```

This is CODE BLOCK 21 in the document R_CODE_BASIC_STATS_WORKBOOK.DOC. Once this command has finished running, you can examine the combined data set created by the `merge` command in the DATA VIEWER window to ensure it contains the data you would expect it to.

At the end of this exercise, the last few lines of your R CONSOLE window should look like this:

```
> names(blue_tit_data)
[1] "box_number" "latitude"   "longitude"  "occupied"   "elevation"
[6] "el_cat"
> names(breeding_data)
[1] "box_number"         "species"            "clutch_size"
[4] "no_chicks_hatched" "no_chicks_fledged"
> combined_dataset <- merge(x=blue_tit_data, y=breeding_data, by="box_number", all.x=TRUE)
> names(blue_tit_data)
[1] "box_number" "latitude"   "longitude"  "occupied"   "elevation"
[6] "el_cat"
> names(breeding_data)
[1] "box_number"         "species"            "clutch_size"
[4] "no_chicks_hatched" "no_chicks_fledged"
> names(combined_dataset)
 [1] "box_number"        "latitude"          "longitude"
 [4] "occupied"          "elevation"         "el_cat"
 [7] "species"           "clutch_size"       "no_chicks_hatched"
[10] "no_chicks_fledged"
> length(blue_tit_data$box_number)
[1] 198
> length(breeding_data$box_number)
[1] 141
> length(combined_dataset$box_number)
[1] 198
> View(combined_dataset)
> |
```

While the contents of the DATA VIEWER window should look like this:

	box_number	latitude	longitude	occupied	elevation	el_cat	species	clutch_size	no_chicks_hatched	no_chicks_fledged
1	3	56.12927	-4.615110	1	32.50000	30 to 40	BT	10	8	8
2	4	56.12952	-4.615516	0	40.00000	30 to 40	NA	NA	NA	NA
3	13	56.13031	-4.616208	0	42.50000	40 to 50	NA	NA	NA	NA
4	14	56.13052	-4.616450	0	40.00000	30 to 40	NA	NA	NA	NA
5	19	56.13106	-4.616647	0	32.61233	30 to 40	GT	6	NA	NA
6	24	56.13129	-4.615251	1	22.87749	20 to 30	BT	12	12	12
7	30	56.13059	-4.614695	0	31.48324	30 to 40	NA	NA	NA	NA
8	33	56.13024	-4.614165	0	30.82850	30 to 40	NA	NA	NA	NA
9	44	56.13033	-4.617238	1	60.00000	50 or more	BT	9	9	9
10	48	56.12906	-4.614811	1	25.00000	20 to 30	BT	12	9	8
11	49	56.12894	-4.615077	0	30.00000	20 to 30	NA	NA	NA	NA
12	66	56.12800	-4.616142	0	34.14250	30 to 40	NA	NA	NA	NA
13	73	56.12795	-4.616873	1	46.66667	40 to 50	BT	10	9	9
14	88	56.12719	-4.616505	0	20.00000	10 to 20	NA	NA	NA	NA
15	89	56.12714	-4.616686	0	15.85750	10 to 20	NA	NA	NA	NA
16	94	56.12737	-4.617562	1	40.00000	30 to 40	BT	7	7	0
17	95	56.12758	-4.617723	0	47.50000	40 to 50	NA	NA	NA	NA
18	96	56.12749	-4.618170	0	45.00000	40 to 50	NA	NA	NA	NA
19	97	56.12695	-4.617856	0	25.00000	20 to 30	NA	NA	NA	NA
20	98	56.12706	-4.618065	0	30.00000	20 to 30	NA	NA	NA	NA

If you wish to change exactly how your two data sets are joined together, you can do this by changing the arguments included in the `merge` command in step 3 of the above flow diagram. Information on how the arguments of the `merge` command can be changed to join data sets together in different ways can be found at *www.rdocumentation.org/packages/base/versions/3.6.1/topics/merge.*

If, once you have joined your data sets together, you wish to export it from R for use in a different software package, you can do this using the `write.table` command. For example, to export the R object called `combined_dataset` created using the `merge` command in the above flow diagram as a .CSV file in your project's WORKING DIRECTORY folder, your `write.table` command would need to look like this:

```
write.table(combined_dataset, file="combined_
                   dataset.csv", sep=",")
```

EXERCISE 1.5: HOW TO CALCULATE SUMMARY STATISTICS FROM A DATA SET IN R:

R can be used to calculate a range of summary statistics from individual data sets. This includes means, medians, standard deviations, maximums, minimums and ranges. These can be calculated individually by running single commands or as a group by running multiple commands at the same time. For this exercise, you will practice calculating summary statistics by generating a table containing eight of the most commonly used ones. These statistics will be calculated from data held in the R object called `blue_tit_data` created in Exercise 1.1 of this workbook, and this means you will need to have completed that exercise before you can start this one. In this exercise, you will use the `subset` command (see Exercise 1.3) to create a new R object containing just the nest boxes occupied by blue tits before your generate the summary tables for these data. **NOTE:** If you wish to generate a set of summary statistics from a whole data set, you do not need to use the `subset` command, and you can simply run the summary statistics code in step 2. The summary statistics you will generate will be based on the land elevation where these boxes are sited and will include the number of boxes occupied by this species, the number of unique elevation values, the minimum elevation, the maximum elevation, the range of elevations (which is the maximum value minus the minimum value), the mean elevation, the median elevation and the standard deviation of elevation. To create this table of summary statistics from the data in the R object called `blue_tit_data`, work though the flow diagram that starts at the top of the next page.

An object in R containing a table of data from which you wish to generate summary statistics

For this example, the R object containing the data you will generate the summary statistics for is called `blue_tit_data` and was created in Exercise 1.1 of this workbook.

1. Identify the name of the column which contains the identifier for the subset of data you will generate your summary statistics for, and the values that are present in it

Before you can extract a subset from your existing data set, you need to check on the names used for the columns in it so you can identify the one you wish to use to create your subset. To do this for the data being used in this example, enter the following command into R:

```
names(blue_tit_data)
```

This code has to be entered exactly as it is written here or it will not work. If you wish to use the copy-and-paste approach for entering this command, copy the text directly below CODE BLOCK 22 in the document R_CODE_ BASIC_STATS_WORKBOOK.DOC and paste it into R. For the data set being used in this example, this will return the names `box_number`, `latitude`, `longitude`, `occupied`, `elevation` and `el_cat`.

Once you have identified the name of the column which you wish to use to create your subset, you need to find out what values it contains. For the data being used in this example you will use the column called `occupied`. To find out what values are contained in this column, enter the following command into R:

```
unique(blue_tit_data$occupied)
```

This is CODE BLOCK 23 in the document R_CODE_ BASIC_STATS_WORKBOOK.DOC. This command will return two values: 1 (which indicates that a specific nest box was occupied by blue tits) and 0 (which indicates it was not occupied by blue tits). **NOTE:** If you get more than two values for this column, or if either of the values returned are not 0 or 1, you will need to return to Exercise 1.1 and re-import the original data set being used in this example.

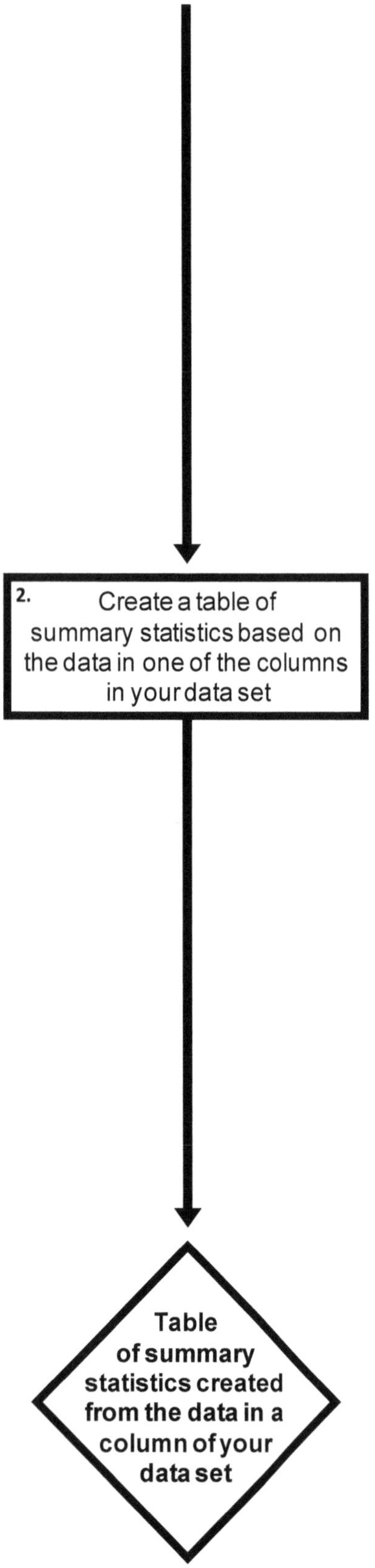

2. Create a table of summary statistics based on the data in one of the columns in your data set

Now that you have identified the column that contains the information that identifies the subset of data you wish to create your summary statistics for (in this case, data with a value of 1 for the column called `occupied`), you need to identify the name of the column that contains the information you wish to summarise. For this example, it will be the column called `elevation` as this contains the land elevation value for each nest box. Once you have this information, you can then enter the code to create your summary statistics table. To do this for the data set being used for this example, enter the following commands into R:

```
occupied_boxes <- subset(blue_tit_data,
occupied=="1")
summary_stats <- data.frame(rbind
(length(occupied_boxes$elevation),
length(unique(occupied_boxes$elevation)),
min(occupied_boxes$elevation),
max(occupied_boxes$elevation),
max(occupied_boxes$elevation)-
min(occupied_boxes$elevation),
mean(occupied_boxes$elevation),
median(occupied_boxes$elevation),
sd(occupied_boxes$elevation)),
row.names=c("Count:","Unique:","Minimum
value:","Maximum value:","Range:","Mean
value:","Median value:","Standard
deviation:"))
colnames(summary_stats)=c("Elevation of
Occupied Boxes")
summary_stats
```

This is CODE BLOCK 24 in the document R_CODE_BASIC_STATS_WORKBOOK.DOC.

Table of summary statistics created from the data in a column of your data set

At the end of this exercise, the last few lines of your R CONSOLE window should look like this:

```
> names(blue_tit_data)
[1] "box_number" "latitude"   "longitude"  "occupied"   "elevation"
[6] "el_cat"
> unique(blue_tit_data$occupied)
[1] 0 1
> occupied_boxes <- subset(blue_tit_data, occupied=="1")
> summary_stats <- data.frame(rbind(length(occupied_boxes$elevation),
+ length(unique(occupied_boxes$elevation)), min(occupied_boxes$elevation),
+ max(occupied_boxes$elevation),
+ max(occupied_boxes$elevation)-min(occupied_boxes$elevation),
+ mean(occupied_boxes$elevation), median(occupied_boxes$elevation), sd(occupied_boxes$elevation)),
+ row.names=c("Count:","Unique values:","Minimum value:","Maximum value:","Range:","Mean value:","Median value:","Standard deviation:"))
> colnames(summary_stats)=c("Elevation of Occupied Boxes")
> summary_stats
                    Elevation of Occupied Boxes
Count:                                 66.00000
Unique values:                         48.00000
Minimum value:                          8.43260
Maximum value:                         60.00000
Range:                                 51.56740
Mean value:                            26.61863
Median value:                          25.00000
Standard deviation:                    11.73875
> |
```

If, once you have created a table of summary statistics for a specific data set, you wish to export it from R for use in a different software package, you can do this using the `write.table` command. For example, to export the R object called `summary_stats` created in this exercise as a .CSV file to your project's WORKING DIRECTORY folder, your `write.table` command would need to look like this:

```
write.table(summary_stats, file="summary_stats.csv",
                           sep=",")
```

As well as the summary statistics generated in this exercise, you can also calculate many other summary statistics. Unfortunately, there isn't a single document which details every possible option you can use. Instead, if you wish to find out how to generate a specific summary statistic other than the ones provided above, the easiest way to do this is to use an internet search engine, such as Google. To do this, enter the name of your desired summary statistic followed by the words FUNCTION IN R. For example, if you wish to calculate a standard error from your data set, you could use the following search term:

Standard error function in R

This should lead you to the R documentation page for the standard error function (which can be found at *www.rdocumentation.org/packages/plotrix/versions/3.7-6/topics/std.error*).

--- Chapter Four ---

Creating Graphs From Biological Data Using R

Once you have successfully imported a data set into R, checked it for errors and organised it, the next thing that you will most likely want to do is to make some graphs from it. Such graphs can serve two important purposes. The first is to provide a visual display of your data for inclusion in presentations and publications, while the second is to help you explore your data so that you can get to know it better. In this chapter, you will learn how to create five types of graphs that biologists commonly need to be able make. These are: 1. Frequency distribution histograms; 2. Bar graphs of count data; 3. Bar graphs of means or medians with error bars; 4. Box plots to show the spread of data in different groups; 5. Scatter plots of the relationships between different variables in a data set. For each one, you will learn about the structure your data need to have to create the graph, as well as all the steps you need to do to make it, starting with a data set held in a spreadsheet or table and finishing with information about how to write a figure legend for it. While this approach of providing complete instructions for each type of graph does result in some repetition of the initial steps between exercises (those for setting the WORKING DIRECTORY, importing data and checking they have been loaded into R correctly), it is designed to make it easier to apply the same workflow to your own data. When working through the exercises in this chapter, you can avoid having to repeat these initial steps, if you have already completed them for a previous exercise, by skipping ahead to the later ones.

If you have not already done so, before you start the exercises in this chapter, you first need to create a WORKING DIRECTORY folder on your computer and load the necessary data into it (**NOTE:** If you have already created this folder and downloaded data for a previous chapter in this workbook, you do not need to do this again). To do this on a computer with a Windows operating system, open Windows Explorer and navigate to the location where you would like to create the folder (such as your C:\ drive or your DOCUMENTS folder). Next, right click anywhere in this location and select NEW> FOLDER. Now call this folder STATS_FOR_BIOLOGISTS_ONE by typing this into the folder name section to replace what it is currently called (which will most likely be NEW FOLDER). To create a WORKING DIRECTORY folder on a computer running a Mac operating system, open Finder and navigate to the location where you would like to create the folder (such as your DOCUMENTS folder or your DESKTOP). Next, click on FILE> NEW FOLDER, and then type the name STATS_FOR_BIOLOGISTS_ONE before pressing the ENTER key on your keyboard.

Once you have created your WORKING DIRECTORY folder, you are ready to download the data sets you will use for the exercises in this workbook from *www.gisinecology.com/stats-for-biologists-1*. After you have downloaded the compressed folder containing the required data by following the instructions provided on that page, you need to extract all the data files from it and copy them into the folder called STATS_FOR_BIOLOGISTS_ONE that you have just created.

Next, you need to check that the required data have been extracted to the correct folder. If you are using a computer with a Windows operating system, you can use Windows Explorer to open your newly created WORKING DIRECTORY folder and examine its contents. If all the files from the compressed folder are present in it (there should be a total of 21 of them), you can click on the folder icon at the left hand end of the ADDRESS BAR at the top of the WINDOWS EXPLORER window to reveal its full address. Write this address down as you will need it to set this folder as your WORKING DIRECTORY during the exercises provided in this workbook (see pages 12 and 13 for details of how to modify folder addresses so they will be recognised by R).

If you are using a computer with a Mac operating system, you can use Finder to open your newly created WORKING DIRECTORY folder and examine its contents. If all the required data files are present in it (there should be a total of 21 of them), select this folder in Finder and then press the CMD and I keys on your keyboard at the same time. This will open the GET INFO window where you will find its address (which is also called the pathway). Write this address down somewhere as you will need it to set this folder as your WORKING DIRECTORY during the exercises provided in this workbook (see pages 12 and 13 for details of how to modify folder addresses so they will be recognised by R).

After you have loaded the required data into your WORKING DIRECTORY folder, you can open RGUI or RStudio, depending on which option you wish to use (see Chapter 2 for more details). Once you have opened your preferred R user interface, you need to create a file called CHAPTER_FOUR_EXERCISES where you will save the results of your analyses from your R CONSOLE window as you work through this chapter. To do this using RGUI, click on the FILE menu and select SAVE WORKSPACE. To do this in RStudio, click on SESSION and select SAVE WORKSPACE AS. In both cases, save it as a WORKSPACE file with the name CHAPTER_FOUR_EXERCISES.RDATA in your WORKING DIRECTORY folder (this will be the one called STATS_FOR_BIOLOGISTS_ONE that you have just created). If you are using RStudio, you will also want to save the contents of your SCRIPT EDITOR window (where you will enter and edit the R code you will use to carry out specific commands). To do this, click on the FILE menu and select SAVE AS. Save your file as an R SCRIPT file with the name CHAPTER_FOUR_EXERCISES.R in your WORKING DIRECTORY folder. As you work through the exercises in this chapter, remember to regularly save the contents of your R CONSOLE window (which will contain the R objects you have created up to that point) to your WORKSPACE file and, if you are using RStudio, the contents of your SCRIPT EDITOR window to your R SCRIPT file.

Finally, you need to remove any data that are currently held in R's temporary memory. To do this, enter the following command into R (if you wish to copy and paste this command, the required code is directly below the text CODE BLOCK 1 in the document called R_CODE_BASIC_STATS_WORKBOOK.DOC that is included in the compressed folder you just downloaded):

```
rm(list=ls())
```

If you are using RGUI, you can simply type or paste this code after the command prompt at the bottom of the R CONSOLE window (it looks like this: >) and then press the ENTER key on your keyboard to run it. If you are using RStudio, you can type or paste this command into the SCRIPT EDITOR window (the upper left hand window). To run this command, select it and then click on the RUN button at the top of this window. This will run it in the R CONSOLE window (the lower left hand one in the main RStudio user interface). You are now ready to start the exercises in this chapter.

EXERCISE 2.1: HOW TO MAKE A FREQUENCY DISTRIBUTION HISTOGRAM:

One of the first graphs that you are likely to want to make is a frequency distribution histogram. This is a graph that shows the distribution of the data in a data set in relation to one of its variables. In order to make a frequency distribution histogram in R, you need to have your data arranged in a spreadsheet or table where each row contains data from a single record in your data set. In this table, there also needs to be a column which contains the values for the continuous variable you wish to create a frequency distribution histogram for. For this exercise, you will start by creating a histogram that shows the distribution of a set of nest boxes used to study breeding behaviour in hole-nesting birds in relation to elevation. To do this, work through the following flow diagram:

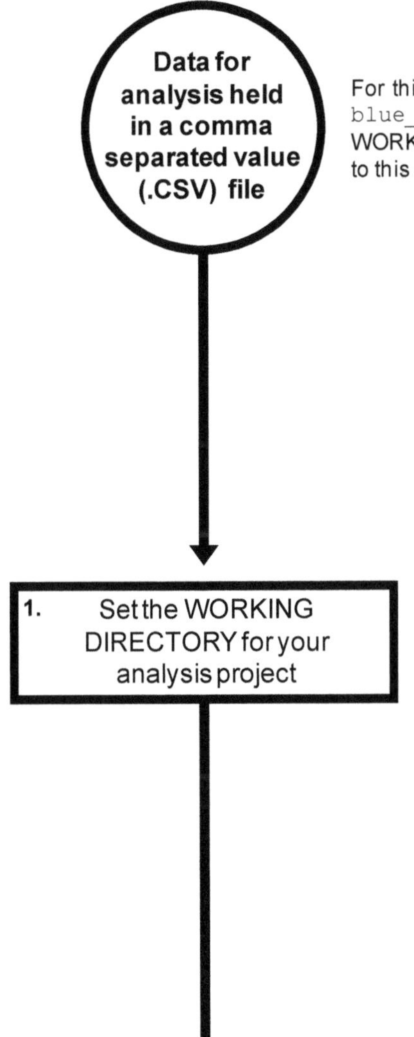

Data for analysis held in a comma separated value (.CSV) file
1. Set the WORKING DIRECTORY for your analysis project

For this example, the data set you will use is stored in a file called `blue_tit_occupancy_data.csv` that is located in the WORKING DIRECTORY folder you created during the introduction to this chapter.

Before you start any analysis in R, you first need to set the WORKING DIRECTORY. To do this, enter the text `setwd("` and then type the address of your WORKING DIRECTORY, using slashes (/) as the folder separators, before entering a second quotation mark followed by a closing bracket, like this `")`. For example, if your WORKING DIRECTORY has the address C:\STATS_FOR_BIOLOGISTS_ONE, your `setwd` command should look like this:

```
setwd("C:/STATS_FOR_BIOLOGISTS_ONE")
```

If you are using RGUI, enter your `setwd` command in the R CONSOLE window (remembering to use the address of your own WORKING DIRECTORY folder in it) and then press the ENTER key on your keyboard. If you are using RStudio, enter your `setwd` command into the SCRIPT EDITOR window. To run it, select it and then click on the RUN button at the top of this window. You will enter all the remaining commands for this exercise in a similar manner, depending on the user interface you are using.

To check that your WORKING DIRECTORY has been set properly, enter the command `getwd()` and carefully check that the address it returns is the same as the one for the STATS_FOR_BIOLOGISTS_ONE folder you created at the start of this chapter.

Before you move on to step 2, make sure that all the data you wish to use in your analysis project are located in this WORKING DIRECTORY folder. In this case, this is a file called `blue_tit_occupancy_data.csv`. **NOTE**: If the data you are going to import into R in step 2 are not located in the WORKING DIRECTORY you set in this step, the import code provided in the next step will not work.

2. Load your data into R using the `read.table` command

The `read.table` command provides the easiest way to load data held in a .CSV file (and stored in the WORKING DIRECTORY you set in step 1) into R so you can analyse it. To do this for the data set being used in this example, enter the following command into R:

```
frequency_histogram_data <- read.table(file=
   "blue_tit_occupancy_data.csv", sep=",",
        as.is=FALSE, header=TRUE)
```

This code has to be entered exactly as it is written here or it will not work. If you wish to use the copy-and-paste approach for entering this command, copy the text directly below CODE BLOCK 25 in the document R_CODE_BASIC_STATS_WORKBOOK.DOC and paste it into R.

This command will create a new object in R called `frequency_histogram_data` which will contain the data from the specified .CSV file. To import a different .CSV file into R, all you need to do is change the file name in the `file` argument to the name of the one you wish to import. You can also use whatever name you wish for the R object which will be created by this command. To do this, simply replace `frequency_histogram_data` at the start of the first line of the above code with the name you wish to use for it. **NOTE:** If your .CSV data set uses a semicolon as the decimal separator, you would need to replace the `sep=","` argument with `sep=";"`.

3. Check the data have loaded into R correctly by checking the names of the columns and by viewing it

Whenever you import any data into R you need to check that they have loaded correctly. First, you need to check that all the required columns are present in the R object you just created. To do this, enter the following command into R:

```
names(frequency_histogram_data)
```

This is CODE BLOCK 26 in the document R_CODE_BASIC_STATS_WORKBOOK.DOC. This command will return the names used for each column in the R object you just created. For this example, the names should be `box_number`, `latitude`, `longitude`, `occupied`, `elevation` and `el_cat`.

Next, you should view the contents of the whole table using the `View` command. This is done by entering following code into R:

```
View(frequency_histogram_data)
```

This is CODE BLOCK 27 in the document R_CODE_BASIC_STATS_WORKBOOK.DOC. This command will open a DATA VIEWER window where you can examine your data set and check that the correct data have been loaded into R.

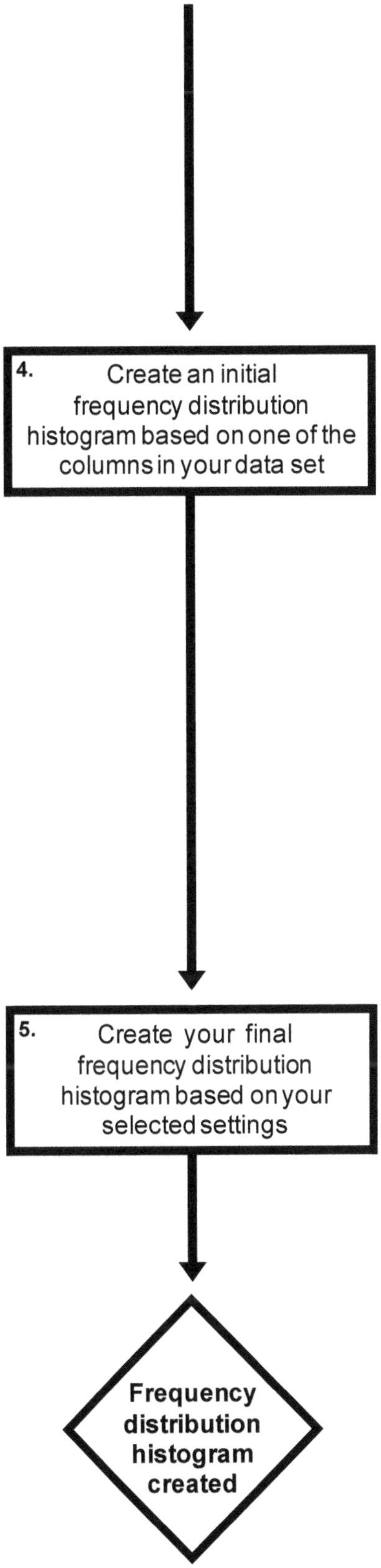

4. Create an initial frequency distribution histogram based on one of the columns in your data set

Once your data have been successfully imported into R, you are ready to create your initial frequency distribution histogram. You will use this initial histogram to check what your final graph will look like. To do this, enter the following command into R:

```
hist(frequency_histogram_data$elevation,
nclass=11)
```

This is CODE BLOCK 28 in the document R_CODE_BASIC_STATS_WORKBOOK.DOC. This `hist` command will create a frequency distribution histogram from the data in the column called `elevation` in the R object called `frequency_histogram_data` created in step 2 of this exercise. The argument `nclass=11` in this command means that the data will be divided into eleven equal-sized classes to make this histogram. If you are not happy with how your frequency distribution histogram looks at this stage, you can try using a different number of classes. This can be done by replacing the `11` in the `nclass` argument with a different number and then re-running it. This can be repeated as often as you need to using different numbers until you are happy with how your frequency distribution histogram looks. **NOTE:** The number of classes in your histogram will not necessarily exactly match the number you have entered in the `nclass` argument. This is because R will adjust the actual number of classes up or down by a small amount from this value to create a histogram with a more appropriate distribution of data between the classes.

5. Create your final frequency distribution histogram based on your selected settings

Once you have found the settings required to make a frequency distribution histogram that you are happy with, you are ready to create your final histogram. This will use the settings for the `hist` command identified in step 4 and include some additional arguments to add labels to it. To do this, enter the following code into R:

```
hist(frequency_histogram_data$elevation,
nclass=11, xlab="Elevation (M)", ylab=
"Number of Nest Boxes", main="Number of Nest
Boxes at Different Elevations")
```

This is CODE BLOCK 29 in the document R_CODE_BASIC_STATS_WORKBOOK.DOC. This code adds three new arguments to the `hist` command. These provide a label for the X axis (specified by the `xlab` argument), a label the Y axis (specified by the `ylab` argument) and a title for the graph (specified by the `main` argument).

Frequency distribution histogram created

At the end of the first part of this exercise, you should have a frequency distribution histogram that looks like this:

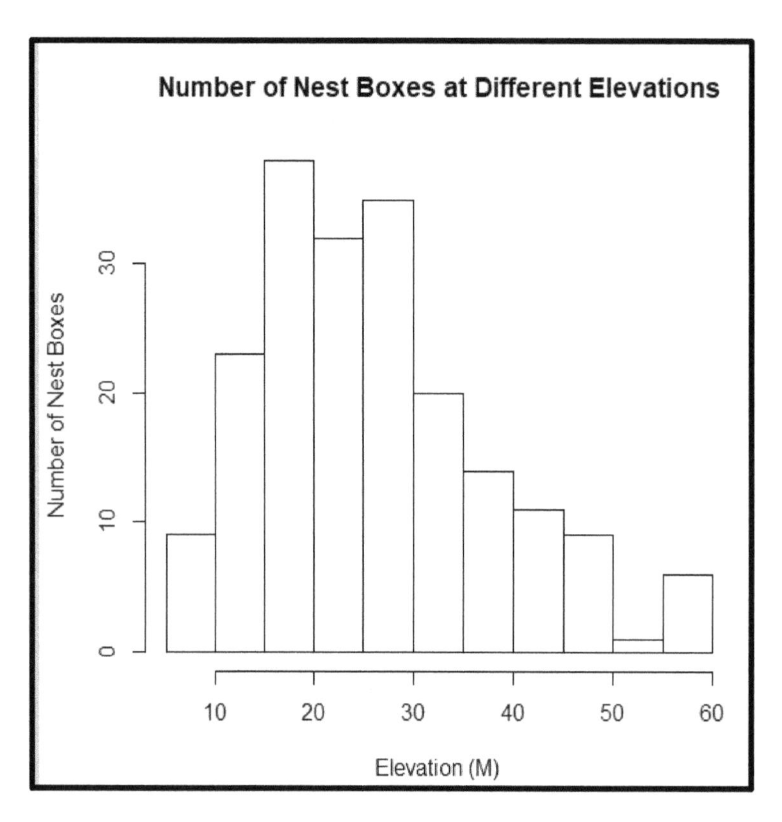

Once you have created a histogram, you can export it from R so that you can include it in a manuscript or presentation. If you are using RGUI, you can do this by clicking on the R GRAPHICS window containing your frequency distribution histogram to select it, before clicking on FILE on the main menu bar and selecting SAVE AS. This will allow you to save it in a variety of different formats. If you are using RStudio, you can export your graph by clicking on the EXPORT button at the top of the window displaying your frequency distribution histogram and selecting SAVE AS IMAGE.

When including a frequency distribution histogram in a manuscript, it is important that you provide an appropriate figure legend for it. This legend should provide all the information required for the reader to interpret the contents of the graph. For the above frequency distribution histogram, an appropriate legend would be:

Figure 1: *The distribution of nest boxes used to study breeding behaviour of hole-nesting birds in relation to the local land elevation (measured in metres).*

The `hist` command used to create the frequency distribution histograms in this exercise can be modified using a range of additional arguments to create a graph with the exact characteristics you wish it to have. The additional arguments that biologists most commonly use when creating histograms are provided in the table below, while the full list of additional arguments that can be used with the `hist` command can be found at *www.rdocumentation.org/packages/graphics/versions/3.6.1/topics/hist*).

Additional Argument	**How To Use It**
`col`	This argument allows you to specify the fill colour of your frequency distribution histogram. The option for this argument can be the name of the desired colour. For example, including the argument `col="blue"` would create a frequency distribution histogram with blue bars on it. Alternatively, you can use a hexadecimal code to specify your desired colour. For example, including the argument `col="#ff3300"` would create a histogram with red bars on it. You can find a full list of hexadecimal codes for different colours at *www.color-hex.com.*
`xlim`	This argument allows you to specify the exact minimum and maximum values for the X axis of your frequency distribution histogram. The options for this argument can be any pair of numbers. For example, including the argument `xlim=c(0,70)` will create a frequency distribution with an X axis that has values ranging from 0 to 70, while including the argument `xlim=c(20,90)` will create a frequency distribution histogram with an X axis that has values ranging from 20 to 90.
`ylim`	This argument allows you to specify the exact minimum and maximum values for the Y axis of your frequency distribution histogram. The options for this argument can be any pair of numbers. For example, including the argument `ylim=c(0,40)` will create a frequency distribution with a Y axis that has values ranging from 0 to 40, while including the argument `ylim=c(20,60)` will create a frequency distribution histogram with a Y axis that has values ranging from 20 to 60.
`main`	This argument allows you to specify the text that will appear as the title of your frequency distribution histogram. The option for this argument is the text you wish to use for your title. For example, including the argument `main="Number of Nest Boxes at Different Elevations"` will create a frequency distribution histogram with this text as the title above it. **<u>NOTE</u>:** This additional argument is optional, and in many cases you will include this information in your figure legend rather than on the graph itself.
`xlab`	This argument allows you to specify the label that will appear alongside the X axis of your frequency distribution histogram. The option for this argument is the text you wish to use as the label for it. For example, including the argument `xlab="Elevation (M)"` will create a frequency distribution histogram with this label alongside its X axis.
`ylab`	This argument allows you to specify the label that will appear alongside the Y axis of your frequency distribution histogram. The option for this argument is the text you wish to use as the label for it. For example, including the argument `ylab="Number of Nest Boxes"` will create a frequency distribution histogram with this label alongside its Y axis.

Additional Argument	How To Use It
labels	This argument allows you to show the number of records in each class above the individual bars of your frequency distribution histogram, or add any other type of label to them. The options for this argument can either be either a list of labels or a logical argument. For example, to add labels that show the number of records represented by each bar on your histogram you would include the additional argument `labels=TRUE`.
nclass	This argument allows you to set the number of bars that will appear on your frequency distribution histogram. The option for this argument can be any positive whole number. The number of classes in your histogram will not necessarily exactly match the number you have entered in the `nclass` argument. This is because R will adjust the actual number of classes up or down from this value to create a histogram with a more appropriate distribution of data between the classes. For example, including the argument `nclass=10` (rather than `nclass=11`) in the above example will still result in a frequency distribution histogram for this data set with 11 bars on it, while including the argument `nclass=5` will create a frequency distribution histogram for this data set with 6 bars on it.

For the next part of this exercise, you will customise your `hist` command in a number of different ways. Firstly, you will alter the contents of the `nclass` argument to change the number of bars (which are referred to as classes in R) used to make your histogram. To do this, change the `nclass` value from 11 to 6 in the `hist` command in step 5 of the above flow diagram. The modified code should look like this (required modifications are highlighted in **bold**):

```
hist(frequency_histogram_data$elevation, nclass=6,
xlab="Elevation (M)", ylab="Number of Nest Boxes",
main="Number of Nest Boxes at Different Elevations")
```

You can modify this code either by editing it in the R CONSOLE window of RGUI or through the SCRIPT EDITOR window of RStudio (depending on which interface you are using). If you are entering commands directly into the R CONSOLE window, you can use the UP arrow on your keyboard to bring commands you have previously run during the same session back on to the command line of this window, and then use the LEFT and RIGHT arrows to scroll through and edit them. In this case, use the UP arrow to bring the previous version of the `hist` command back onto the command line and edit it so that it looks like the one above. Once you have finished modifying this command, you can run it by pressing the ENTER key on your keyboard. If you are using RStudio, you can copy and paste the original `hist` command in the SCRIPT EDITOR window before editing the new version to include the required modifications. Once you have done this, you can select the modified version of the command and click on the RUN button to run it in the R CONSOLE window. This modified code should produce a new histogram that looks like this:

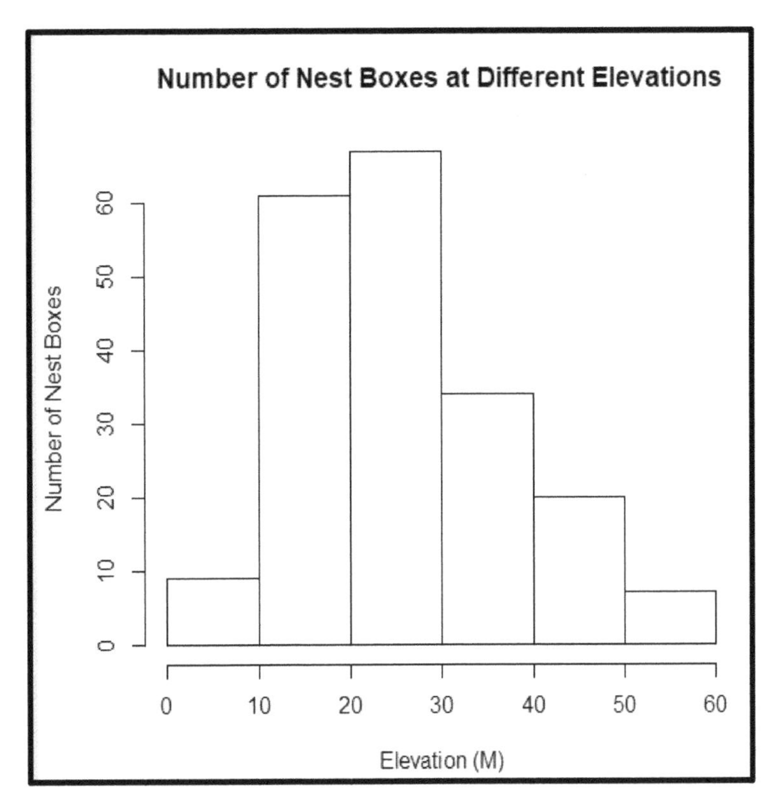

Next, you are going to further customise your `hist` command by adding a new additional argument, `col`, to it. This additional argument allows you to set the fill colour for the bars of your histogram. For this exercise, you will now add the argument `col="blue"` to the above `hist` command. The modified code should look like this:

```
hist(frequency_histogram_data$elevation, nclass=6,
xlab="Elevation (M)", ylab="Number of Nest Boxes",
main="Number of Nest Boxes at Different Elevations",
col="blue")
```

When you run this version of the `hist` command, it should produce a new histogram that looks like the image at the top of the next page (with the bars coloured blue rather than being unfilled).

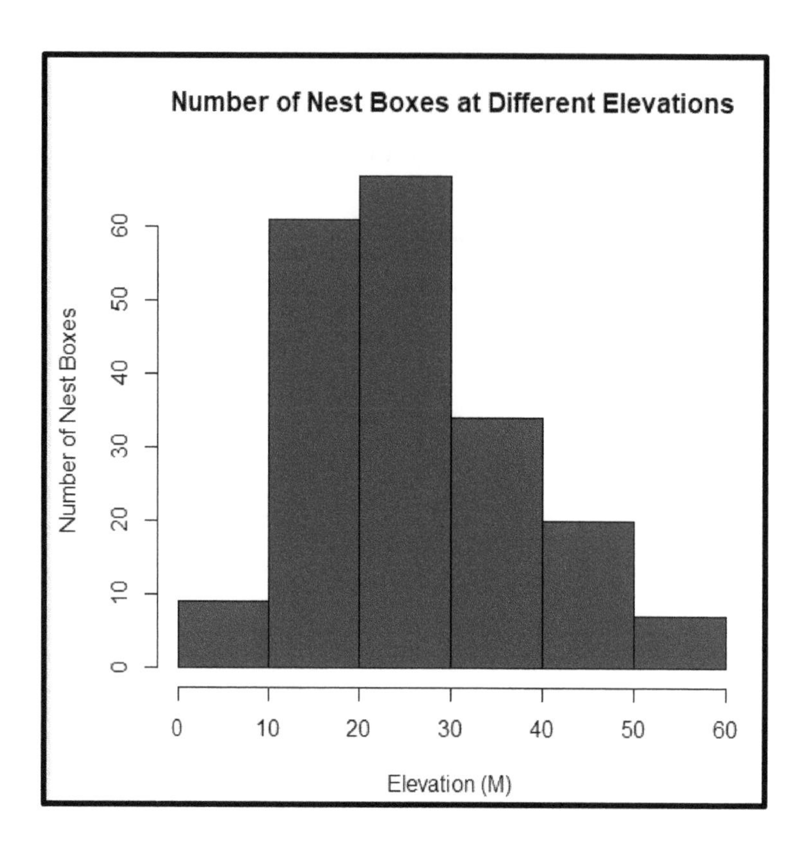

Once you have learned more about how to use the hist command by customising it, you can test your ability to use it correctly by entering the following variation on the original hist command into R:

```
hist(frequency_histogram_data$elevation, nclass=11,
xlab="Elevation (M)", ylab="Number of Nest Boxes",
main="Number of Nest Boxes at Different Elevations",
col="red", xlim=c(0,70), ylim=c(0,40))
```

This command will create a histogram with 11 classes on it (set by the nclass argument) and made of red bars (set by the col argument). In addition, two new additional arguments have been added to set the range of values that are displayed on the X axis (using the xlim additional argument) and on the Y axis (using the ylim additional argument). When you run this version of the hist command, it should produce a new histogram that looks like the image at the top of the next page.

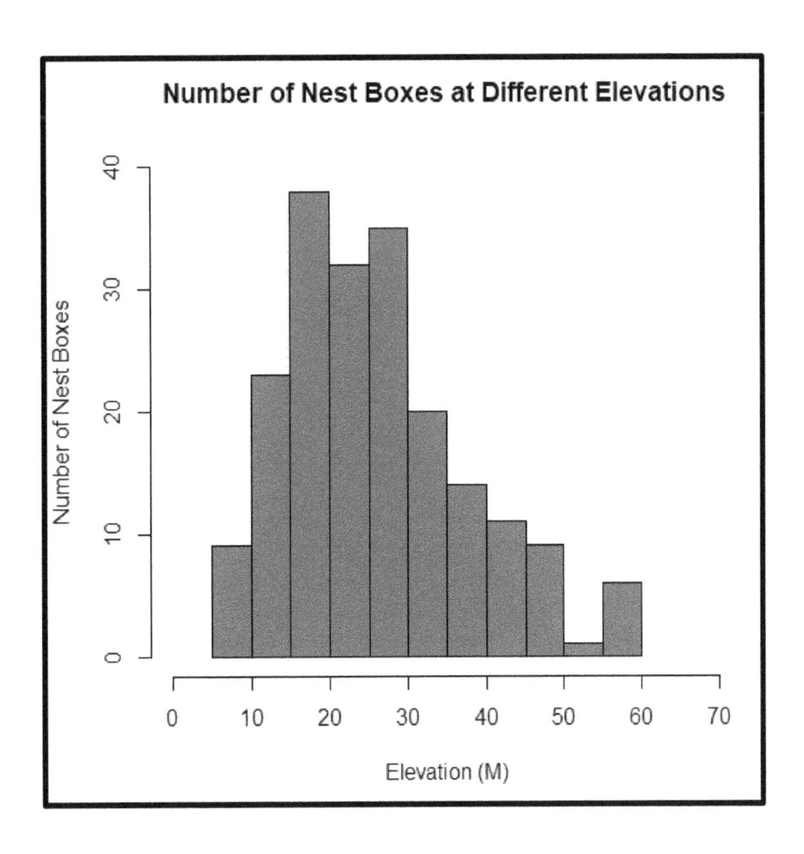

EXERCISE 2.2: HOW TO MAKE A BAR GRAPH BASED ON THE NUMBER OF RECORDS IN DIFFERENT CATEGORIES IN A DATA SET:

Histograms are generally used to show the distribution of data within a data set in relation to a continuous variable (such as elevation in Exercise 2.1). While bar graphs look superficially similar to histograms, they differ in one key way: they display data in relation to categorical variables rather than continuous variables. There are two types of data that biologists commonly display on bar graphs. These are count data and summary statistics (such as the mean values of a specific variable for all records in each group in a data set – see Exercise 2.3). In order to create a bar graph that displays count data in R, you will need to have your data arranged in a spreadsheet or table where each row contains data from a single record in your data set. In this table, there also needs to be a column that contains the categorical variable that will be used for the X axis of your bar graph. For this exercise, you will start by creating a bar graph that shows the number of nest boxes sited in different elevation categories that were occupied by blue tits during one breeding season. The Y axis of this graph will show the count of the number of records in each elevation category based on a column that contains a 1 to indicate blue tit presence and a 0 to indicate absence. To do this, work through the following flow diagram:

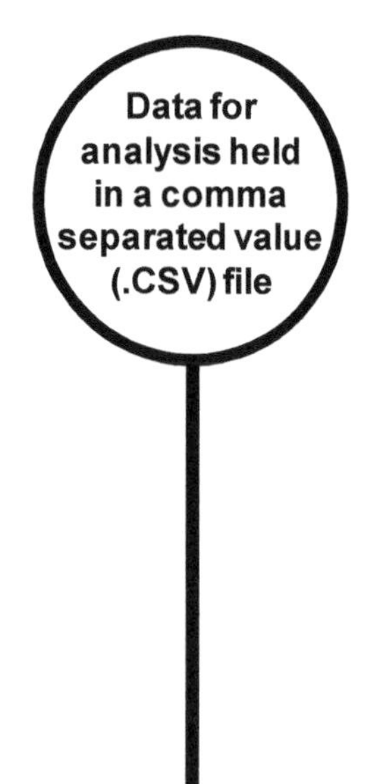

For this example, the data set you will use is stored in a file called `blue_tit_occupancy_data.csv` that is located in the WORKING DIRECTORY folder you created during the introduction to this chapter.

1. Set the WORKING DIRECTORY for your analysis project

Before you start any analysis in R, you first need to set the WORKING DIRECTORY. To do this, enter the text `setwd("` and then type the address of your WORKING DIRECTORY, using slashes (/) as the folder separators, before entering a second quotation mark followed by a closing bracket, like this `")`. For example, if your WORKING DIRECTORY has the address C:\STATS_FOR_BIOLOGISTS_ONE, your `setwd` command should look like this:

```
setwd("C:/STATS_FOR_BIOLOGISTS_ONE")
```

If you are using RGUI, enter your `setwd` command in the R CONSOLE window (remembering to use the address of your own WORKING DIRECTORY folder in it) and then press the ENTER key on your keyboard. If you are using RStudio, enter your `setwd` command into the SCRIPT EDITOR window. To run it, select it and then click on the RUN button at the top of this window. You will enter all the remaining commands for this exercise in a similar manner, depending on the user interface you are using.

To check that your WORKING DIRECTORY has been set properly, enter the command `getwd()` and carefully check that the address it returns is the same as the one for the STATS_FOR_BIOLOGISTS_ONE folder you created at the start of this chapter.

Before you move on to step 2, make sure that all the data you wish to use in your analysis project are located in this WORKING DIRECTORY folder. In this case, this is a file called `blue_tit_occupancy_data.csv`. **NOTE**: If the data you are going to import into R in step 2 are not located in the WORKING DIRECTORY you set in this step, the import code provided in the next step will not work.

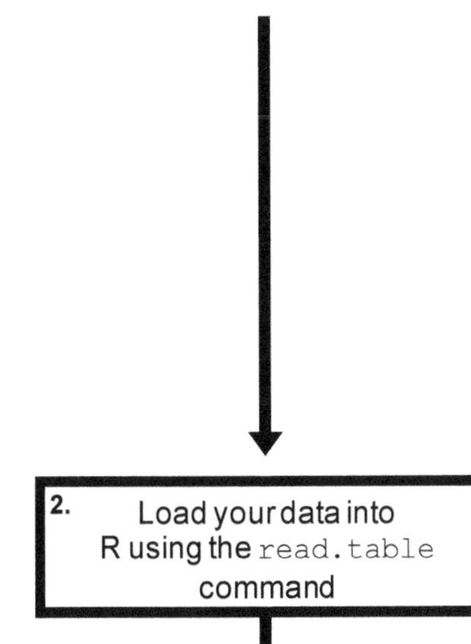

2. Load your data into
R using the read.table
command

The `read.table` command provides the easiest way to load data held in a .CSV file (and stored in the WORKING DIRECTORY you set in step 1) into R so you can analyse it. To do this for the data set being used in this example, enter the following command into R:

```
blue_tit_data <- read.table(file="blue_tit_
occupancy_data.csv", sep=",", as.is=FALSE,
header=TRUE)
```

This code has to be entered exactly as it is written here or it will not work. If you wish to use the copy-and-paste approach for entering this command, copy the text directly below CODE BLOCK 30 in the document R_CODE_BASIC_STATS_WORKBOOK.DOC and paste it into R.

This command will create a new object in R called `blue_tit_data` which will contain the data from the specified .CSV file. To import a different .CSV file into R, all you need to do is change the file name in the `file` argument to the name of the one you wish to import. You can also use whatever name you wish for the R object which will be created by this command. To do this, simply replace `blue_tit_data` at the start of the first line of the above code with the name you wish to use for it. **NOTE:** If your .CSV data set uses a semicolon as the decimal separator, you would need to replace the `sep=","` argument with `sep=";"`.

3. Check the data have loaded into R correctly by checking the names of the columns and by viewing it

Whenever you import any data into R you need to check that they have loaded correctly. First, you need to check that all the required columns are present in the R object you just created. To do this, enter the following command into R:

```
names(blue_tit_data)
```

This is CODE BLOCK 31 in the document R_CODE_BASIC_STATS_WORKBOOK.DOC. This command will return the names used for each column in the R object created in step 2. For this example, the names should be `box_number`, `latitude`, `longitude`, `occupied`, `elevation` and `el_cat`.

Next, you should view the contents of the whole table using the `View` command. This is done by entering the following code into R:

```
View(blue_tit_data)
```

This is CODE BLOCK 32 in the document R_CODE_BASIC_STATS_WORKBOOK.DOC. This command will open a DATA VIEWER window where you can examine your data set and check that the correct data have been loaded into R.

4. Create a new table from your data set containing the information you wish to display on your bar graph

Since a bar graph displays information based on categories, you will need to create a new table from your data set which contains this information and save it as a new R object in your analysis project. In this example, you wish to display the number of nest boxes occupied by blue tits in the different elevation categories provided in the column called `el_cat`. To do this, you first need to create a new table from your data using the `aggregate` command by entering the following code into R:

```
barplot_data <- aggregate(blue_tit_data$
occupied, list(blue_tit_data$el_cat), sum)
colnames(barplot_data)=c("el_cat",
"occupied")
```

This is CODE BLOCK 33 in the document R_CODE_BASIC_STATS_WORKBOOK.DOC. These commands will create a new R object called `barplot_data` consisting of two columns. The first will be called `el_cat` and will contain a list of the different categories in the `el_cat` column of the R object called `blue_tit_data`. The second will be called `occupied` and will contain the total number of occupied nest boxes in each elevation category. This total is calculated by the `sum` argument which adds up all the values in the `occupied` column of the original `blue_tit_data` R object for each elevation category. This column contains a 1 for any nest boxes that were occupied by blue tits and a 0 for any that were not. This means that the sum of values for this column for each elevation category will equal the number of occupied nest boxes within it. **NOTE:** If you just wanted to plot the total number of nest boxes in each category, you would use the `length` argument rather than the `sum` argument at the end of `aggregate` command.

5. Create an initial bar graph based on the summary table created in step 4

Once you have created your summary table as a new object in R, you are ready to create your initial bar graph. You will use this initial bar graph to check what your final graph will look like. To do this, enter the following command into R:

```
barplot(barplot_data$occupied)
```

This is CODE BLOCK 34 in the document R_CODE_BASIC_STATS_WORKBOOK.DOC. This command will create a bar graph from the data in the column called `occupied` in the R object called `barplot_data` created in step 4. On this bar graph, each bar will represent a count of the number of nest boxes occupied by blue tits in each elevation category in the initial data set. At this stage, you can review your graph to ensure that it is showing exactly the information you wish it to show. If it doesn't, you would need to go back and repeat steps 4 and 5 to ensure that your summary table contains the correct information (step 4) and that you are using the correct column from it to create your bar graph (this step).

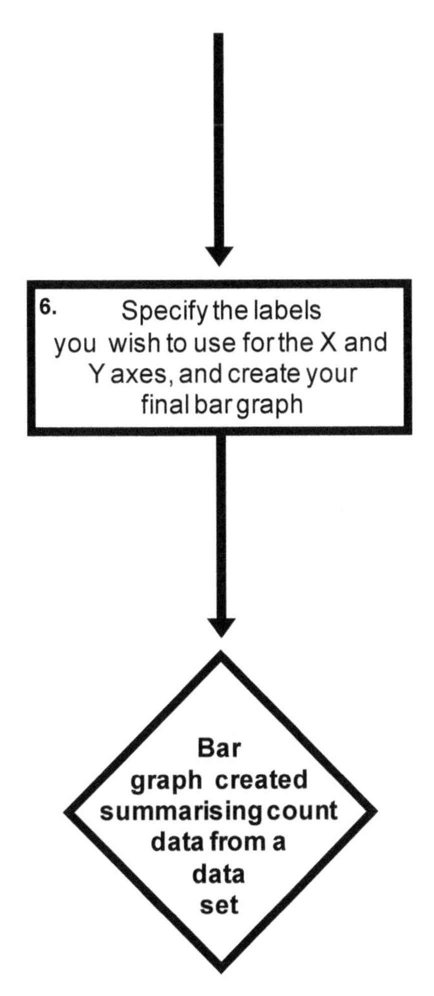

6. Specify the labels you wish to use for the X and Y axes, and create your final bar graph
Bar graph created summarising count data from a data set

Once you have checked your initial graph and you are happy that it shows the required information, you are ready to create your final bar graph. To do this, enter the following command into R:

```
barplot(barplot_data$occupied,
xlab="Elevation Categories", ylab="Number of
Occupied Boxes", names.arg=barplot_
data$el_cat)
```

This is CODE BLOCK 35 in the document R_CODE_BASIC_STATS_WORKBOOK.DOC. This code adds three new arguments to the `barplot` command. These provide a label for the X axis (specified by the `xlab` argument), a label for the Y axis (specified by the `ylab` argument) and a label for the categories shown on the X axis (specified by the `names.arg` argument).

Once you have worked through the first part of this exercise, you need to check the contents of the summary table (called `barplot_data`) created in step 4. If it is not already visible, use the command `View(barplot_data)` to open a DATA VIEWER window so that you can see it. Its contents should look like this:

	el_cat	occupied
1	0 to 10	3
2	10 to 20	20
3	20 to 30	23
4	30 to 40	11
5	40 to 50	7
6	50 or more	2

The bar graph that you created from this table in step 6 should look like this:

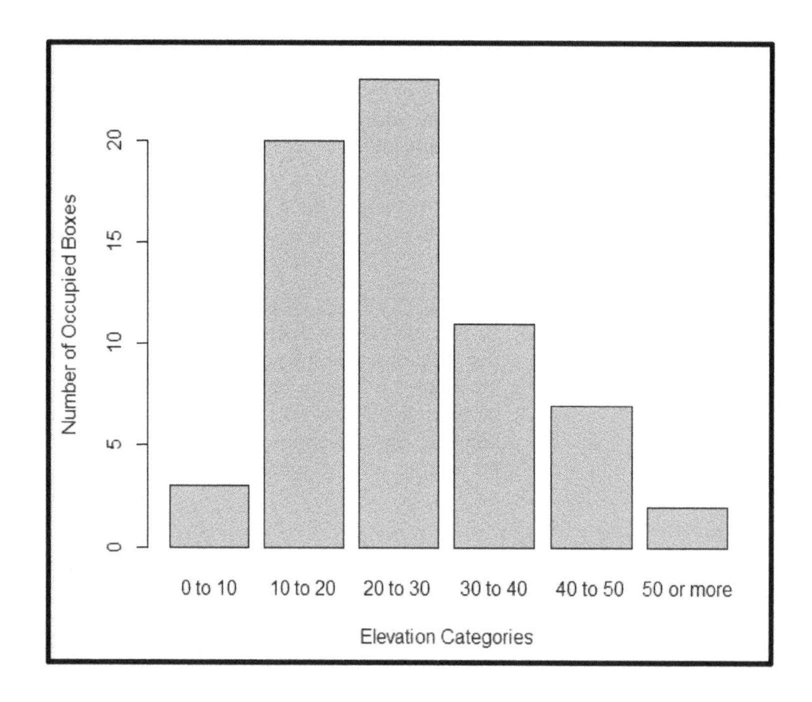

Once you have created a bar graph, you can export it from R so that you can include it in a manuscript or presentation. If you are using RGUI, you can do this by clicking on the R GRAPHICS window containing your bar graph to select it, before clicking on FILE on the main menu bar and selecting SAVE AS. This will allow you to save it in a variety of different formats. If you are using RStudio, you can export your graph by clicking on the EXPORT button at the top of the window displaying your bar graph and selecting SAVE AS IMAGE.

When including a bar graph in a manuscript, it is important that you provide an appropriate figure legend for it. This legend should provide all the information required for the reader to interpret the contents of the graph. For the above bar graph, an appropriate legend would be:

Figure 1: *The number of nest boxes occupied by blue tits during a single breeding season in six elevation categories (in metres).*

The `barplot` command used to create bar graphs in this exercise can be modified using a range of additional arguments to create a graph with the exact characteristics you wish it to have. The additional arguments that biologists most commonly use when creating such graphs are provided in the table below, while the full list of additional arguments that can be used with the `barplot` command can be found at *www.rdocumentation.org/packages/graphics/versions/3.6.1/topics/barplot*).

Additional Argument	How To Use It
`col`	This argument allows you to specify the fill colour of the bars of your bar graph. The option for this argument can be the name of the desired colour. For example, including the argument `col="blue"` would create a bar graph with blue bars on it. Alternatively, you can use a hexadecimal code to specify your desired colour. For example, including the argument `col="#ff3300"` would create a bar graph with red bars on it. You can find a full list of hexadecimal codes for different colours at *www.color-hex.com.*
`beside`	This argument allows you to specify whether the values for different bars on your graph should be drawn beside each other or stacked on top of each other. The options for this argument are `TRUE` or `FALSE`. Using the argument `beside=TRUE` will create a bar graph where all the bars are drawn side by side. In contrast, using the arguments `beside=FALSE` will create a bar graph where the bars are stacked on top of each other.
`ylim`	This argument allows you to specify the minimum and maximum values for the Y axis of your bar graph. The options for this argument can be any pair of numbers. For example, including the argument `ylim=c(0,40)` would create a bar graph with a Y axis that has values ranging from 0 to 40, while including the argument `ylim=c(20,60)` would create a bar graph with a Y axis that has values ranging from 20 to 60.
`names.arg`	This argument allows you to specify the names that will be displayed on the X axis below the bar for each group. The option that you will use most often for this argument is the name of a column in the data set used to make your bar graph. For example, including the argument `names.arg=barplot_data$el_cat` would create a graph where the labels displayed below the bars along the X axis are obtained from a column called `el_cat` in an R object called `barplot_data`.
`main`	This argument allows you to specify the text that will appear as the title of your bar graph. The option for this argument is the text you wish to use for your title. For example, including the argument `main="Number of Nest Boxes at Different Elevations"` will create a bar graph with this text as the title above it. **NOTE:** This additional argument is optional, and in many cases you will include this information in your figure legend rather than on the graph itself.
`legend.text`	This argument allows you to add a simple legend to your graph to identify which data sets the bars on it represent. The easiest way to use this command is to specify the names you wish to have on your legend. For example, including the argument `legend.text= C("Blue Tits","Great Tits")` will create a legend using these names. To add more complex legends, use the `legend` command instead (see *www.rdocumentation.org/packages/graphics/versions/3.6.1/topics/legend* for more details).

Additional Argument	How To Use It
`xlab`	This argument allows you to specify the label that will appear alongside the X axis of your bar graph. The option for this argument is the text you wish to use as the label for it. For example, including the argument `xlab="Elevation Categories"` will create a bar graph with this label alongside its X axis.
`ylab`	This argument allows you to specify the label that will appear alongside the Y axis of your bar graph. The option for this argument is the text you wish to use as the label for it. For example, including the argument `ylab="Number of Occupied Boxes"` will create a bar graph with this label alongside its Y axis.
`width`	This argument allows you to specify the width of each bar on your bar graph. The options for this argument are numeric values indicating the required width for each bar. As a result, there should be the same number of values in this argument as there are bars on your graph. For example, for the bar graph created from the above flow diagram, you would need to specify six values in this argument. Using the argument `width=c(100,100,100,100,100,50)` for a bar graph with six bars would set the first five bars to a width of 100, and the final bar to half this width.
`space`	This argument allows you to specify the gap between each bar on your bar graph. The option for this argument is a numeric value indicating the fraction of the bar width left before each bar. It can be a single number, or a list of numbers if you wish to have different gaps between each bar. For example, using the argument `space=0` would create a bar graph with no gaps between the bars, while using the argument `space=0.1` would create a bar graph where the gap was 10% of the width of the bars.

For the next part of this exercise, you will customise your `barplot` command in a number of different ways. Firstly, you will change the fill colour of the bars on your bar graph to blue. To do this, add the additional argument `col="blue"` to the `barplot` command in step 6 of the above flow diagram. The modified code to do this should look like this (required modifications are highlighted in **bold**):

```
barplot(barplot_data$occupied, xlab="Elevation
Categories ylab="Number of Occupied Boxes", names.arg=
barplot_data$el_cat, col="blue")
```

You can modify this code either by editing it in the R CONSOLE window of RGUI or through the SCRIPT EDITOR window of RStudio (depending on which interface you are using). If you are entering commands directly into the R CONSOLE window, you can use the UP arrow on your keyboard to bring commands you have previously run during the same session back on to the command line of this window, and then use the LEFT and RIGHT arrows to scroll through and edit them. In this case, use the UP arrow to bring the previous version of the barplot command back onto the command line and edit it so that it looks like the one above. Once you have finished modifying your barplot command, you can run it by pressing the ENTER key on your keyboard. If you are using RStudio, you can copy and paste the original barplot command in the SCRIPT EDITOR window before editing the new version to include the required modifications. Once you have done this, you can select the modified version of the command and click on the RUN button to run it in the R CONSOLE window. This modified code should produce a new bar graph that looks like this:

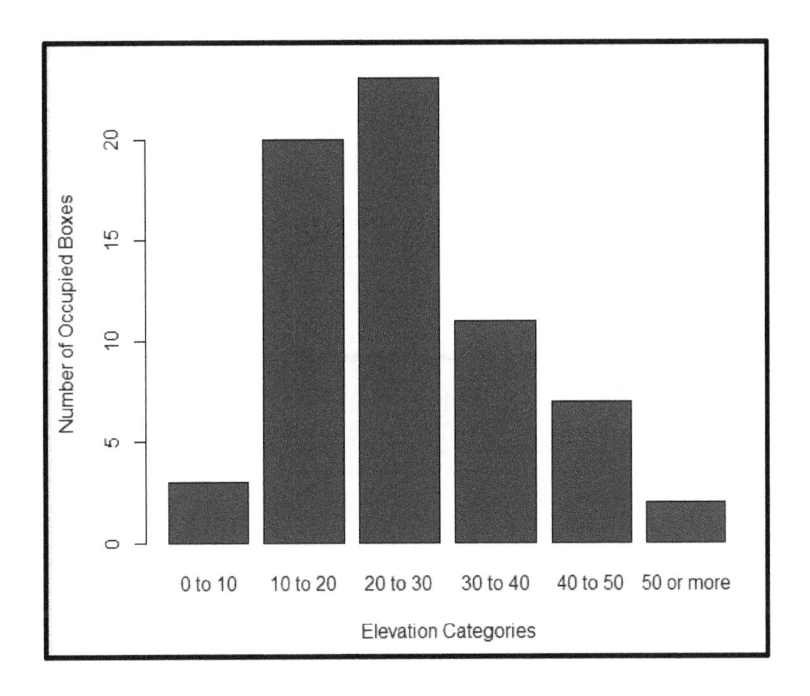

You can also customise the barplot command to display two series of data on the same graph. However, before you can do this, you would need to create an individual summary table for each data series (following the instructions in step 4), and then use the rbind argument to plot both data series on the same graph. To practice doing this, you first need to modify the command in step 2 of the above flow diagram to import a data set that identifies which nest boxes from the same study were occupied by a second species of hole-nesting bird, the great tit. This information is contained in a file called great_tit_occupancy_data.csv and it should be located in the WORKING DIRECTORY you set in step 1. The modified read.table command required to import this new data set should look like the code provided at the top of the next page.

```
great_tit_data <- read.table(file="great_tit_occupancy_
    data.csv", sep=",", as.is=FALSE, header=TRUE)
```

You can then create a summary table based on this newly imported data set containing the same columns as you used for the blue tit data by modifying the R code provided in step 4 of the above flow diagram. The modified code to do this should look like this:

```
barplot_data_2 <- aggregate(great_tit_data$occupied,
    list(great_tit_data$el_cat),sum)
colnames(barplot_data_2)=c("el_cat","occupied")
```

You can check the table created by this modified code using the `View` command by entering the following code into R:

```
View(barplot_data_2)
```

It should look like this:

	el_cat	occupied
1	0 to 10	2
2	10 to 20	9
3	20 to 30	9
4	30 to 40	2
5	40 to 50	5
6	50 or more	1

After this second summary table has been created, you can include the `rbind` argument in the `barplot` command to create a bar graph displaying the summary information from both the original blue tit summary table (called `barplot_data`) and the new great tit summary table you just created (called `barplot_data_2`). At this stage, you can also set the colours that you will use for each data series by specifying two different colours (in this case, `blue` and `red`) in the `col` argument. The modified code to do this should look like this:

```
barplot(rbind(barplot_data$occupied, barplot_
data_2$occupied), beside=TRUE, xlab="Elevation Categories",
    ylab="Number of Occupied Boxes", names.arg=barplot_
    data$el_cat, col=c("blue","red"))
```

When you run this version of the `barplot` command, it should produce a new bar graph that looks like this:

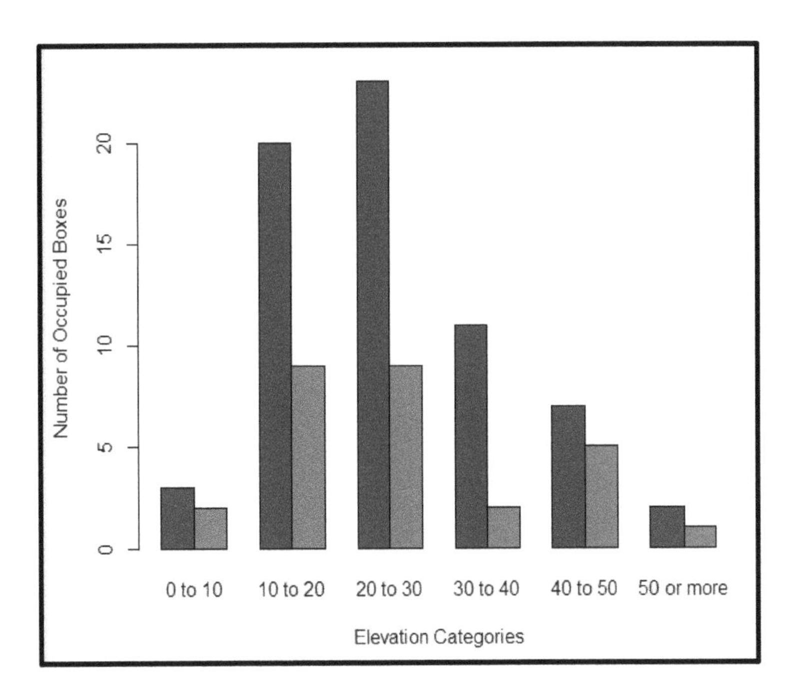

When creating a figure legend for graphs with more than one data series on it, you need to make it clear which data are represented by which colour of bars. For the above bar graph, an appropriate legend would be:

Figure 1: *The number of nest boxes occupied by blue tits (blue) and great tits (red) in six elevation categories (in metres) in a single breeding season.*

Once you have learned more about how to use the `barplot` command by customising it, you can test your ability to use it correctly by entering the following command into R:

```
barplot(rbind(barplot_data$occupied, barplot_
data_2$occupied), beside=TRUE, xlab="Elevation Categories",
ylab="Number of Occupied Boxes", names.arg=barplot_
data$el_cat, col=c("blue","red"), ylim=c(0,25), legend.
text=c("Blue Tits","Great Tits"))
```

This version of the `barplot` command will create a bar graph that is identical to the previous graph, with two exceptions. Two new argument have been added, `ylim`, which sets the range of values displayed on the Y axis to those between 0 and 25, and `legend.text`, which adds a legend to the graph. When you run this version of the `barplot` command, it should produce a bar graph that looks like this:

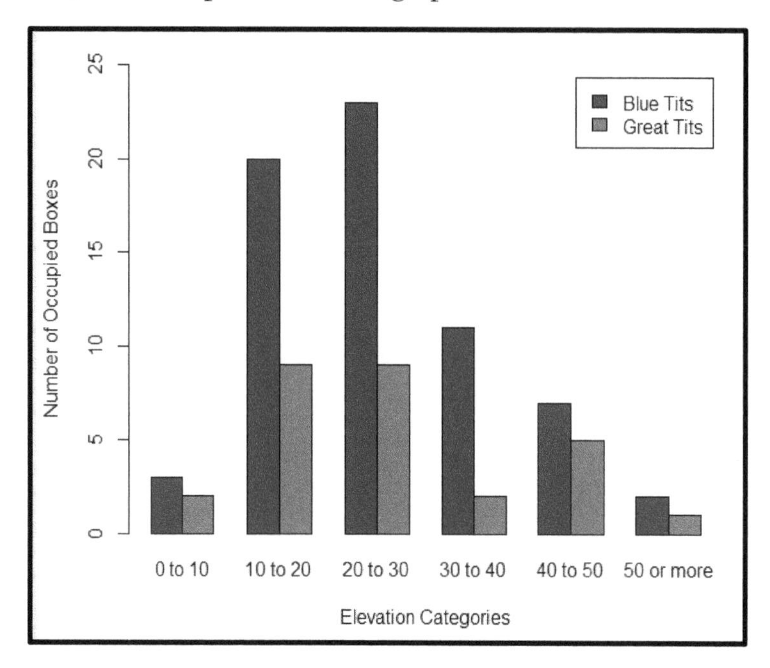

If you compare this to the previous version of this graph, you will be able to see the difference in the range of values displayed on the Y axis, and the impact this has on the height of the different bars shown on it, as well as the legend that has been added to it.

EXERCISE 2.3: HOW TO MAKE A BAR GRAPH OF MEAN OR MEDIAN VALUES WITH ERROR BARS:

While bar graphs are commonly used to display count data (see Exercise 2.2), they can also be used to display summary statistics for a data set. For example, bar graphs can be used to display mean or median values for a specific variable to allow you to compare these values for different categories of a second variable in a data set, such as species or sampling location. When plotting such graphs, you will usually need to add error bars to show the variations around the mean or median values they represent. In order to be able to do this in R, you will need to have your data arranged in a spreadsheet or table where each row contains data from a single record in your data set. In this table, there also needs to be a column containing the variable that will be used for the categories on the X axis of your bar graph, and a second column containing the continuous variable which will be used to calculate the mean/median values for each category, and the measure of variation to be used for the error bars, that will be plotted on the Y axis. For this exercise, you will start by creating a bar graph that shows the mean clutch size for blue tits from a data set which has information about breeding success (including clutch size, number of hatchlings and number of fledglings) from the same nest boxes used for earlier exercises in this workbook. The instructions for how to do this are provided in the following flow diagram (**NOTE:** This information is held in a different data set from the occupancy data used for the first two exercises in this chapter):

For this example, the data set you will use is stored in a file called `blue_tit_breeding_data.csv` that is located in the WORKING DIRECTORY folder you created during the introduction to this chapter.

1. Set the WORKING DIRECTORY for your analysis project

Before you start any analysis in R, you first need to set the WORKING DIRECTORY. To do this, enter the text `setwd("` and then type the address of your WORKING DIRECTORY, using slashes (/) as the folder separators, before entering a second quotation mark followed by a closing bracket, like this `")`. For example, if your WORKING DIRECTORY has the address C:\STATS_FOR_BIOLOGISTS_ONE, your `setwd` command should look like this:

```
setwd("C:/STATS_FOR_BIOLOGISTS_ONE")
```

If you are using RGUI, enter your `setwd` command in the R CONSOLE window (remembering to use the address of your own WORKING DIRECTORY folder in it) and then press the ENTER key on your keyboard. If you are using RStudio, enter your `setwd` command into the SCRIPT EDITOR window. To run it, select it and then click on the RUN button at the top of this window. You will enter all the remaining commands for this exercise in a similar manner, depending on the user interface you are using.

To check that your WORKING DIRECTORY has been set properly, enter the command `getwd()` and carefully check that the address it returns is the same as the one for the STATS_FOR_BIOLOGISTS_ONE folder you created at the start of this chapter.

Before you move on to step 2, make sure that all the data you wish to use in your analysis project are located in this WORKING DIRECTORY folder. In this case, this is a file called `blue_tit_breeding_data.csv`. **NOTE:** If the data you are going to import into R in step 2 are not located in the WORKING DIRECTORY you set in this step, the import code provided in the next step will not work.

2. Load your data into R using the `read.table` command

The `read.table` command provides the easiest way to load data held in a .CSV file (and stored in the WORKING DIRECTORY you set in step 1) into R so you can analyse it. To do this for the data set being used in this example, enter the following command into R:

```
blue_tit_breeding_data <- read.table(file=
  "blue_tit_breeding_data.csv", sep=",",
        as.is=FALSE, header=TRUE)
```

This code has to be entered exactly as it is written here or it will not work. If you wish to use the copy-and-paste approach for entering this command, copy the text directly below CODE BLOCK 36 in the document R_CODE_BASIC_STATS_WORKBOOK.DOC and paste it into R.

This command will create a new object in R called `blue_tit_breeding_data` which will contain the data from the specified .CSV file. To load a different .CSV file into R, all you need to do is change the file name in the `file` argument to the name of the one you wish to import. In addition, you can use whatever name you wish for the R object which will be created by this command. To do this, simply replace `blue_tit_breeding_data` at the start of the first line of the above code with the name you wish to use for it. **NOTE:** If your .CSV data set uses a semicolon as the decimal separator, you would need to replace the `sep=","` argument with `sep=";"`.

3. Check the data have loaded into R correctly by checking the names of the columns and by viewing it

Whenever you import any data into R, you need to check that they have loaded correctly. First, you need to check that all the required columns are present in the R object you just created. To do this, enter the following command into R:

```
names(blue_tit_breeding_data)
```

This is CODE BLOCK 37 in the document R_CODE_BASIC_STATS_WORKBOOK.DOC. This command will return the names used for each column in the R object created in step 2. For this example, the names should be `box_number`, `elevation`, `el_cat`, `species`, `clutch_size`, `no_chicks_hatched` and `no_chicks_fledged`.

Next, you should view the contents of the whole table using the `View` command. This is done by entering the following code into R:

```
View(blue_tit_breeding_data)
```

This is CODE BLOCK 38 in the document R_CODE_BASIC_STATS_WORKBOOK.DOC. This command will open a DATA VIEWER window where you can examine your data set and check that the correct data have been loaded into R.

4. Create a table from your data set containing the information you wish to display on your bar graph

Since a bar graph displays information based on categories, you will need to create a summary table from your data set which contains this information and save it as a new R object in your analysis project. In this example, you will create a bar graph that will display the mean clutch size in nest boxes occupied by blue tits in different elevation categories provided in a column called `el_cat`. Before you can do this, you first need to create a new table from your data using the `aggregate` command by entering the following code into R:

```
mean <- aggregate(blue_tit_breeding_
data$clutch_size, list(blue_tit_breeding_
data$el_cat), mean)
colnames(mean)=c("el_cat","mean")
```

This is CODE BLOCK 39 in the document R_CODE_BASIC_STATS_WORKBOOK.DOC. These commands will create a new R object called `mean` consisting of two columns. The first will be called `el_cat` and will contain a list of the different categories in the `el_cat` column of the `blue_tit_breeding_data` data set. The second will be called `mean` and it will contain the mean clutch size in each elevation. These values are calculated using the `mean` argument in the `aggregate` command.

5. Create a table from your data set containing the information you wish use for your error bars

Next, you need to calculate the metric you wish to use for your error bars. In this example, you will calculate the standard deviation. To do this, you will create a second table using the `aggregate` command by entering the following code into R:

```
st_dev <- aggregate(blue_tit_breeding_
data$clutch_size, list(blue_tit_breeding_
data$el_cat), sd)
colnames(st_dev)=c("el_cat","st_dev")
```

This is CODE BLOCK 40 in the document R_CODE_BASIC_STATS_WORKBOOK.DOC. These commands will create a new R object called `st_dev` consisting of two columns. The first will be called `el_cat` and will contain a list of the different categories in the `el_cat` column of the `blue_tit_breeding_data` data set. The second will be called `st_dev` and it will contain the standard deviation of clutch size in each elevation category. These values are calculated using the `sd` argument in the `aggregate` command.

6. Join together the tables containing the information for your bar graph and your error bars

Once you have created the separate tables with the summary statistics for your bar graph and your error bars, you need to join the two tables together. To do this, you will use the `merge` command by entering the following code into R:

```
mean_and_st_dev <- merge(mean,st_dev,
               by="el_cat")
colnames(mean_and_st_dev)=c("el_cat","mean",
               "st_dev")
```

This is CODE BLOCK 41 in the document R_CODE_BASIC_STATS_WORKBOOK.DOC. These commands will create a new table in a new R object called `mean_and_st_dev` by joining together the data from the tables created in step 4 (`mean`) and step 5 (`st_dev`) based on the shared `el_cat` column. Once you have created this new table, you should examine it to make sure the numbers in it look appropriate. This can be done by entering the following command into R:

```
View(mean_and_st_dev)
```

This is CODE BLOCK 42 in the document R_CODE_BASIC_STATS_WORKBOOK.DOC.

7. Create your bar graph based on the summary table created in step 6

Once you have created your combined table, you are ready to create your bar graph. To do this, enter the following command into R:

```
bar_graph <- barplot(height=mean_and_
  st_dev$mean, names.arg=mean_and_
  st_dev$el_cat, ylim=c(0,15), xlab=
"Elevation Categories", ylab="Clutch Size",
 main="Blue Tit Clutch Size by Elevation")
```

This is CODE BLOCK 43 in the document R_CODE_BASIC_STATS_WORKBOOK.DOC. This command creates a new object in R called `bar_graph` which contains the graph created by the `barplot` command. The height of each bar on this graph is defined by the values in the `mean` column of the `mean_and_st_dev` R object created in step 6. This means that the bar heights will display the mean clutch size for blue tits in each individual elevation category in this table.

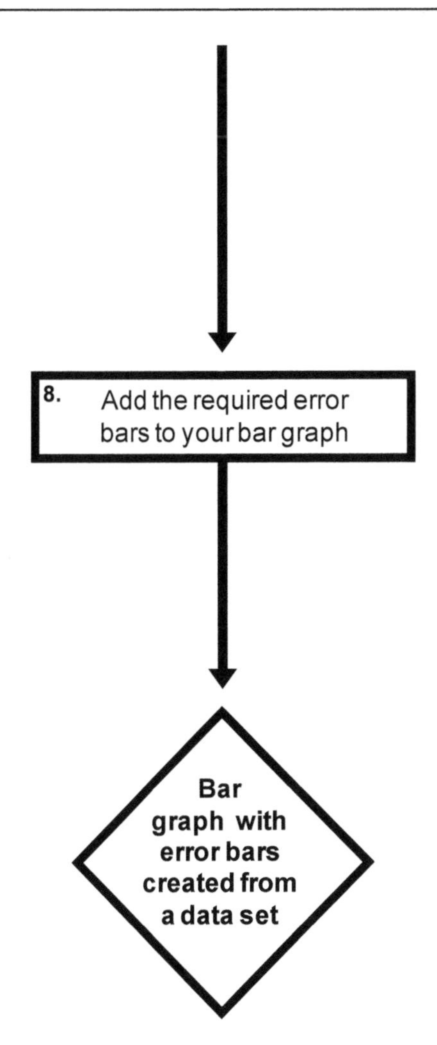
8. Add the required error bars to your bar graph
Bar graph with error bars created from a data set

Once you have created the R object containing your bar graph, you can then add error bars to it using the `segments` command. To do this for the data being used in this example, enter the following code into R:

```
segments(bar_graph, mean_and_st_
dev$mean - mean_and_st_dev$st_dev,
bar_graph, mean_and_st_dev$mean + mean_
and_st_dev$st_dev, lwd = 2)
```

This is CODE BLOCK 44 in the document R_CODE_BASIC_STATS_WORKBOOK.DOC. This code uses the `segments` command to add a vertical line above and below the bar for each mean value. The length of these lines will be the standard deviation values for each category calculated in step 5. These line are plotted by subtracting the standard deviation value from the mean value (for the line coming downwards) and by adding the standard deviation value to the mean value (for the line reaching upwards). The width of these lines is set by the `lwd` argument.

Once you have worked through this example, you should check the contents of the summary table created in step 6. If it is not already visible, use the command `View(mean_and_st_dev)` to open a DATA VIEWER window so that you can see it. It should look like this:

	el_cat	mean	st_dev
1	0 to 10	9.666667	2.516611
2	10 to 20	8.294118	1.896204
3	20 to 30	8.047619	1.986862
4	30 to 40	8.142857	1.345185
5	40 to 50	8.400000	1.516575

The bar graph with error bars that you created from this table in steps 7 and 8 should look like the image at the top of the next page.

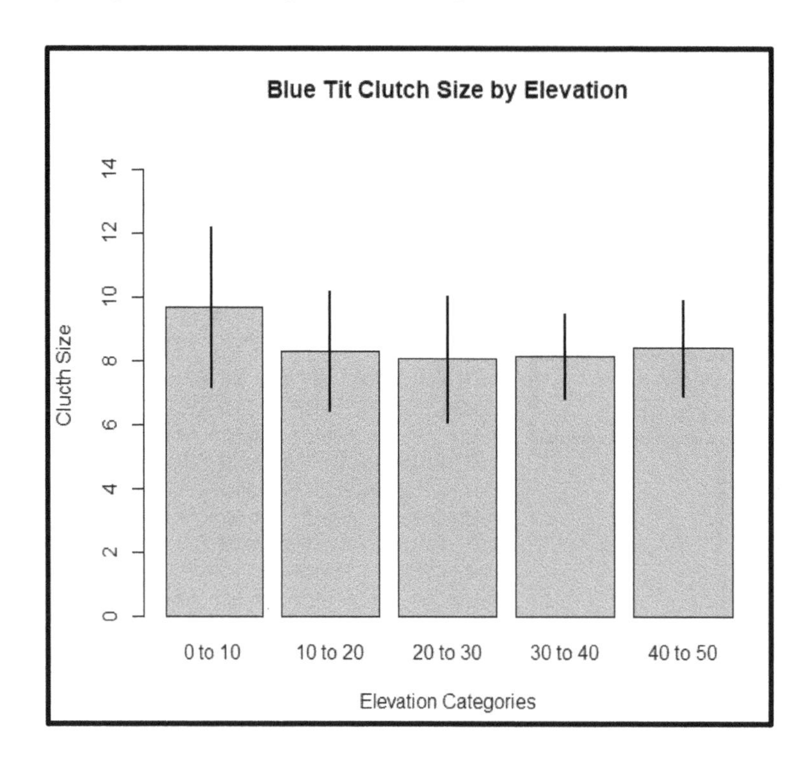

Once you have created a bar graph with error bars, you can export it from R so that you can include it in a manuscript or presentation. If you are using RGUI, you can do this by clicking on the R GRAPHICS window containing bar graph to select it, before clicking on FILE on the main menu bar and selecting SAVE AS. This will allow you to save it in a variety of different formats. If you are using RStudio, you can export your graph by clicking on the EXPORT button at the top of the window displaying your bar graph and selecting SAVE AS IMAGE.

When including a bar graph with mean or median values and error bars in a manuscript, it is important that you provide an appropriate figure legend for it. This legend should provide all the information required for the reader to interpret the contents of the graph. For above the bar graph, an appropriate legend would be:

Figure 1: *The clutch size of blue tits in nest boxes in six elevation categories. The bars represent the mean clutch size recorded in each elevation category, while the error bars show the standard deviation of clutch sizes recorded in that category.*

This method for plotting a bar graph showing mean or median values and adding error bars to it can be modified in a number of different ways. For example, in the above graph the error bars are displayed as simple vertical lines. However, you can also add a short horizontal bar to the ends of the error bars to provide a clearer indication of their upper and lower limits. This is done by changing the `segments` command to the `arrows` command as outline below and adding it as a new line of code below the original `segments` command (required modifications are highlighted in **bold**).

```
arrows(bar_graph, mean_and_st_dev$mean - mean_and_st_dev
$st_dev, bar_graph, mean_and_st_dev$mean + mean_and_st_dev
$st_dev, lwd=2, angle=90, code=3, length=0.1)
```

This new `arrows` command will add a horizontal line to the top and bottom of your existing error bars on your existing bar graph (saved in the R object called `bar_graph`), and it should now look like this:

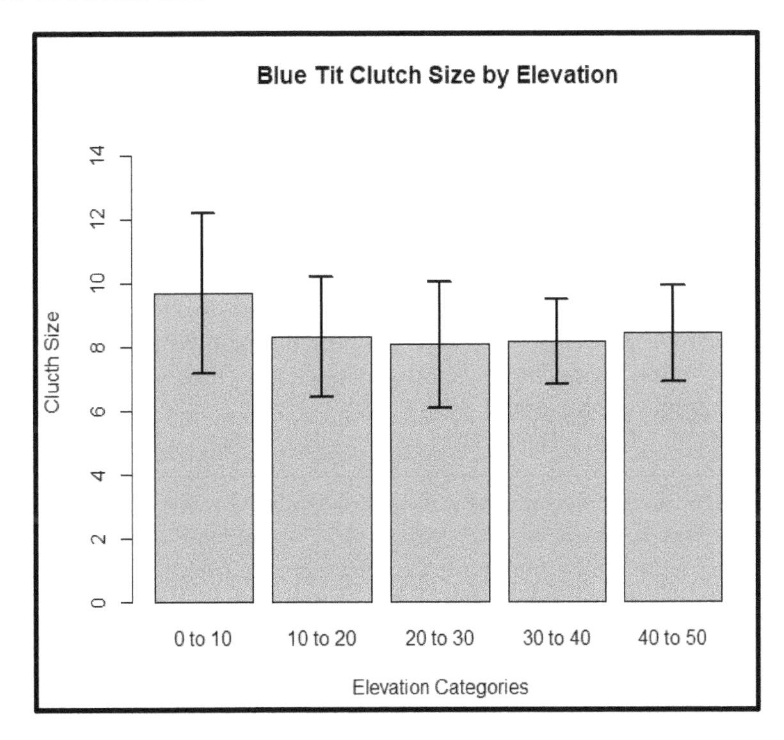

As well as modifying the commands associated with adding error bars to your bar graph, you can also modify whether the graph displays the mean or the median values and what the error bars represent. This is done by changing the information calculated when making the summary tables created in steps 4 and 5. The table on the next page provides information about the arguments you can use as part of the `aggregate` command to generate different summary statistics that you can use to plot different types of information on a bar graph.

Argument	How To Use It
mean	This argument allows you to calculate the mean value for each group of data. To calculate the mean for individual groups within a data set, use the argument `mean` at the end of the `aggregate` command in step 4 of the above flow diagram.
median	This argument allows you to calculate the median value for each group of data. To calculate the median value for individual groups within a data set, use the argument `median` at the end of the `aggregate` command in step 4 of the above flow diagram.
sd	This argument allows you to calculate the standard deviation for each group of data. To calculate the standard deviation for individual groups within a data set, use the argument `sd` at the end of the `aggregate` command in step 5 of the above flow diagram.
std.error	This argument allows you to calculate the standard error for each group of data. In order to be able to use this argument in the `aggregate` command, you first need to install the Plotrix package. This can be done using the `install.packages ("plotrix")` command, followed by the `library(plotrix)` command. To calculate the standard error for individual groups within a data set, use the argument `std.error` at the end of the `aggregate` command in step 5 of the above flow diagram.
IQR	This argument allows you to calculate the interquartile range for each group of data. To calculate the interquartile range for individual groups within a data set, use the argument `IQR` at the end of the `aggregate` command in step 5 of the above flow diagram. <u>**NOTE**</u>: In order to plot error bars representing the interquartile range, you will need to divide this number by 2 to get the appropriate length of the error bars above and below the median value. This can be calculated and added to the table in step 5 of the above flow diagram by adding the following line of code after the `aggregate` and the `colnames` commands (this code assumes that you have used the name IQR for the summary table containing this information, and that you have labelled the column containing this information IQR): `IQR$HALF_IQR=(IQR$IQR/2)`
group.CI	A confidence interval cannot easily be calculated using the `aggregate` command. Instead, you need to use the `group.CI` command in place of the `aggregate` command in step 5 of the above flow diagram. To use this command, you will first need to install the Rmisc package. This can be done using the `install.packages ("Rmisc")` command, followed by the `library(Rmisc)` command. The options for the `group.CI` command are the column containing the data you wish to generate a confidence interval from followed by the column with the groups you wish to calculate it for, the R object that contains these data, and the required confidence interval. For example, using the command `group.CI(clutch_size~el_cat, data= blue_tit_breeding_data, 0.95)` will calculate a table with the 95% confidence intervals for a column called `clutch_size` based on the groups in a column called `el_cat` in an R object called `blue_tit_breeding_data`. <u>**NOTE:**</u> To add these confidence intervals as error bars to your graph, you will need to modify the `segments` command in step 8 so that it uses the correct data for them. These data are held in the columns labelled `clutch_size.upper` and `clutch_size.lower` in the R object created by this command.

For the next part of this exercise, you will change the information displayed by the error bars from the standard deviation of clutch size to the standard error. To do this, you will need to modify step 5 so that rather than creating a table containing the standard deviations of the clutch sizes recorded in each elevation category, it will create a table of standard errors. This can be done using the following modified version of the `aggregate` command:

```
se <- aggregate(blue_tit_breeding_data$clutch_size,
  list(blue_tit_breeding_data$el_cat),std.error)
colnames(se)=c("el_cat","se")
View(se)
```

If you get an error message when you run this modified command, this will most likely be due to the library which contains the `std.error` command (called Plotrix) not being loaded into R. To load this library into your version of R, enter the following code and then follow any instructions that appear:

```
install.packages("plotrix")
```

Once this package has been installed, you can load the required library by entering the following command:

```
library(plotrix)
```

You can now re-run the above `aggregate` command to create your table of standard errors (which will be called `se`). The resulting summary table should look like this (if it is not already visible, use the `View(se)` command to open it in a DATA VIEWER window):

	el_cat	se
1	0 to 10	1.4529663
2	10 to 20	0.4598969
3	20 to 30	0.4335687
4	30 to 40	0.5084323
5	40 to 50	0.6782330

Once this standard error table has been created, join it to the table of means (called `mean`) by modifying the code provided in step 6 of the above flow diagram so it looks like this:

```
mean_and_se <- merge(mean,se,by="el_cat")
colnames(mean_and_se)=c("el_cat","mean","se")
View(mean_and_se)
```

At this point, you are ready to re-make the bar graph showing the mean values by following step 7 (by re-running the `barplot` command at this stage, you will remove the error bars that are currently displayed on it), and then add error bars showing standard error (rather than standard deviations) to this graph by modifying the code provided in step 8 as follows:

```
segments(bar_graph, mean_and_se$mean - mean_and_se$se, bar_
graph, mean_and_se$mean + mean_and_se$se, lwd=2)
arrows(bar_graph, mean_and_se$mean - mean_and_se$se, bar_
graph, mean_and_se$mean + mean_and_se$se, lwd=2,
angle=90, code=3, length=0.1)
```

If you apply these modifications to the above flow diagram and work through it, the resulting bar graph should look like this:

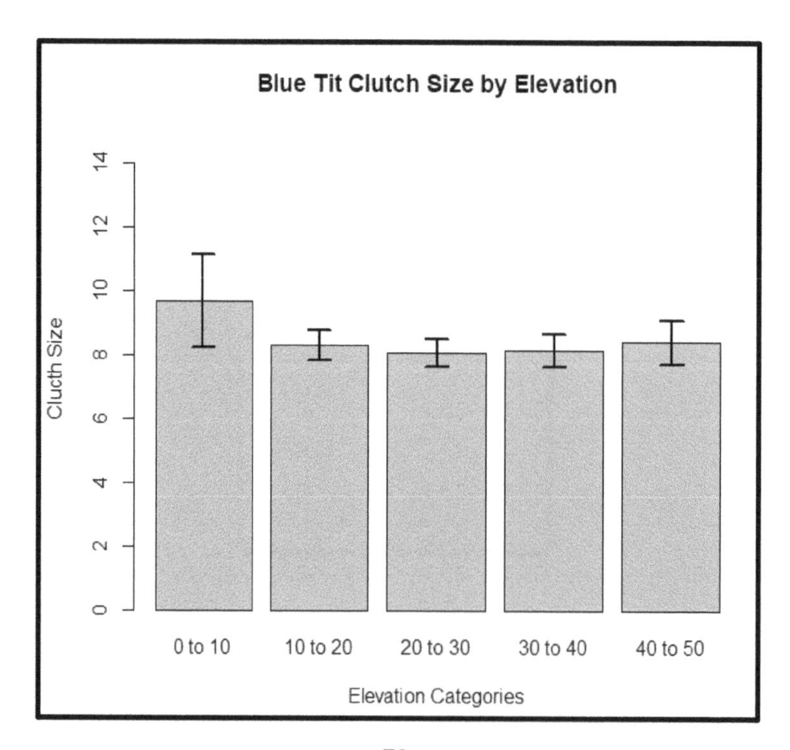

Once you have learned more about how to create a bar graph of central values and add error bars to a it, you can test your ability to do this correctly by creating a plot showing the mean number of fledglings in each elevation category, complete with error bars showing the standard error, by entering the following code (representing modified versions of the code provided in steps 4 to 8 of the above flow diagram) into R.

Modified Code Required For Step 4:

```
mean_2 <- aggregate(blue_tit_breeding_data$no_chicks_
fledged, list(blue_tit_breeding_data$el_cat), mean)
colnames(mean_2)=c("el_cat","mean")
```

Modified Code Required For Step 5:

```
se_2 <- aggregate(blue_tit_breeding_data$no_chicks_fledged,
list(blue_tit_breeding_data$el_cat), std.error)
colnames(se_2)=c("el_cat","se")
```

Modified Code Required For Step 6:

```
mean_and_se_2 <- merge(mean_2, se_2, by="el_cat")
colnames(mean_and_se_2)=c("el_cat","mean","se")
View(mean_and_se_2)
```

Modified Code Required For Step 7:

```
bar_graph <- barplot(height= mean_and_se_2$mean, names.arg=
mean_and_se_2$el_cat, ylim=c(0,15), xlab="Elevation
Categories", main="Blue Tit Fledging Success by Elevation",
ylab="No. Fledged")
```

Modified Code Required For Step 8:

```
segments(bar_graph, mean_and_se_2$mean - mean_and_se_2$se,
bar_graph, mean_and_se_2$mean + mean_and_se_2$se, lwd=2)
arrows(bar_graph, mean_and_se_2$mean - mean_and_se_2$se,
bar_graph, mean_and_se_2$mean + mean_and_se_2$se, lwd=2,
angle=90, code=3, length=0.1)
```

Once you have entered and run this modified code, it should produce a new bar graph that looks like this:

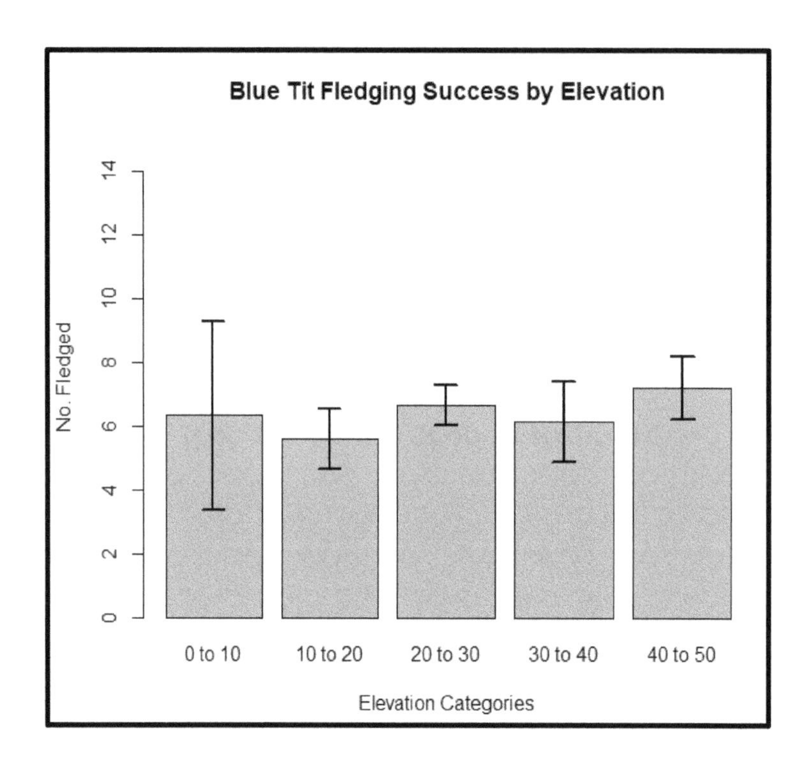

If your graph does not look like this, use the `View` command to make sure that the summary table generated by the modified commands steps 4 to 6 (called `mean_and_se_2`) has been created properly. It should look like this:

	el_cat	mean	se
1	0 to 10	6.333333	2.9627315
2	10 to 20	5.588235	0.9277535
3	20 to 30	6.666667	0.6261497
4	30 to 40	6.142857	1.2426656
5	40 to 50	7.200000	0.9695360

EXERCISE 2.4: HOW TO MAKE A BOX PLOT TO SHOW THE SPREAD OF DATA WITHIN DIFFERENT GROUPS IN A DATA SET:

Box plots are commonly used to explore data sets and carry out preliminary comparisons of the range of values for a particular variable found in different groups in a data set. For example, box plots can be used to explore whether there are differences in the range of values for a particular variable between seasons, between areas, between a control and an experimental set-up or between different species. Such graphical comparisons can also be used to help determine which statistical tests might be suitable for testing whether or not any observed differences are significant. For example, if three groups appear to differ in their mean values, you might want to use an ANOVA to investigate whether the differences are significant or not. However, if they appear to contain different ranges of values, you will first need to do an F-test or Levene's test to assess if these differences are significant. If there are significant differences in the variances between the groups of data, this is inconsistent with the requirements of an ANOVA, and you would need to use the non-parametric Kruskal-Wallis test instead (see Chapter Six for more information on these tests, including how to use them).

In order to be able to create a box plot in R, you will need to have your data arranged in a spreadsheet or table where each row contains data from a single record in your data set. In this table, there also needs to be one column containing a value for the continuous variable you wish to show the ranges for on the Y axis, and a second column containing a categorical variable that you will use to identify which group each record belongs to. For this exercise, you will start by creating a box plot that compares the clutch size of two species of hole-nesting bird, the blue tit and the great tit. To create this graph, work through the flow diagram that starts at the top of the next page.

Data for analysis held in a comma separated value (.CSV) file

For this example, the data set you will use is stored in a file called `nestbox_breeding_data.csv` that is located in the WORKING DIRECTORY folder you created during the introduction to this chapter.

1. Set the WORKING DIRECTORY for your analysis project

Before you start any analysis in R, you first need to set the WORKING DIRECTORY. To do this, enter the text `setwd("` and then type the address of your WORKING DIRECTORY, using slashes (/) as the folder separators, before entering a second quotation mark followed by a closing bracket, like this `")`. For example, if your WORKING DIRECTORY has the address C:\STATS_FOR_BIOLOGISTS_ONE, your `setwd` command should look like this:

```
setwd("C:/STATS_FOR_BIOLOGISTS_ONE")
```

If you are using RGUI, enter your `setwd` command in the R CONSOLE window (remembering to use the address of your own WORKING DIRECTORY folder in it) and then press the ENTER key on your keyboard. If you are using RStudio, enter your `setwd` command into the SCRIPT EDITOR window. To run it, select it and then click on the RUN button at the top of this window. You will enter all the remaining commands for this exercise in a similar manner, depending on the user interface you are using.

To check that your WORKING DIRECTORY has been set properly, enter the command `getwd()` and carefully check that the address it returns is the same as the one for the STATS_FOR_BIOLOGISTS_ONE folder you created at the start of this chapter.

Before you move on to step 2, make sure that all the data you wish to use in your analysis project are located in this WORKING DIRECTORY folder. In this case, this is a file called `nestbox_breeding_data.csv`. **NOTE:** If the data you are going to import into R in step 2 are not located in the WORKING DIRECTORY you set in this step, the import code provided in the next step will not work.

2. Load your data into R using the read.table command
3. Check the data have loaded into R correctly by checking the names of the columns and by viewing it

The `read.table` command provides the easiest way to load data held in a .CSV file (and stored in the WORKING DIRECTORY you set in step 1) into R so you can analyse it. To do this for the data set being used in this example, enter the following command into R:

```
boxplot_data <- read.table(file="nestbox_
breeding_data.csv", sep=",", as.is=FALSE,
            header=TRUE)
```

This code has to be entered exactly as it is written here or it will not work. If you wish to use the copy-and-paste approach for entering this command, copy the text directly below CODE BLOCK 45 in the document R_CODE_ BASIC_STATS_WORKBOOK.DOC and paste it into R.

This command will create a new object in R called `boxplot_data` which will contain the data from the specified .CSV file. To load a different .CSV file into R, all you need to do is change the file name in the `file` argument to the name of the one you wish to import. In addition, you can use whatever name you wish for the R object which will be created by this command. To do this, simply replace `boxplot_data` at the start of the first line of the above code with the name you wish to use for it. **NOTE:** If your .CSV data set uses a semicolon as the decimal separator, you would need to replace the `sep=","` argument with `sep=";"`.

Whenever you import any data into R, you need to check that they have loaded correctly. First, you need to check that all the required columns are present in the R object you just created. To do this, enter the following command into R:

```
names(boxplot_data)
```

This is CODE BLOCK 46 in the document R_CODE_ BASIC_STATS_WORKBOOK.DOC. This command will return the names of each column in the R object created in step 2. For this example, the names should be `box_number`, `species`, `clutch_size`, `no_chicks_ hatched` and `no_chicks_fledged`.

Next, you should view the contents of the whole table using the `View` command. This is done by entering the following code into R:

```
View(boxplot_data)
```

This is CODE BLOCK 47 in the document R_CODE_ BASIC_STATS_WORKBOOK.DOC. This command will open a DATA VIEWER window where you can examine your data set and check that the correct data have been loaded into R.

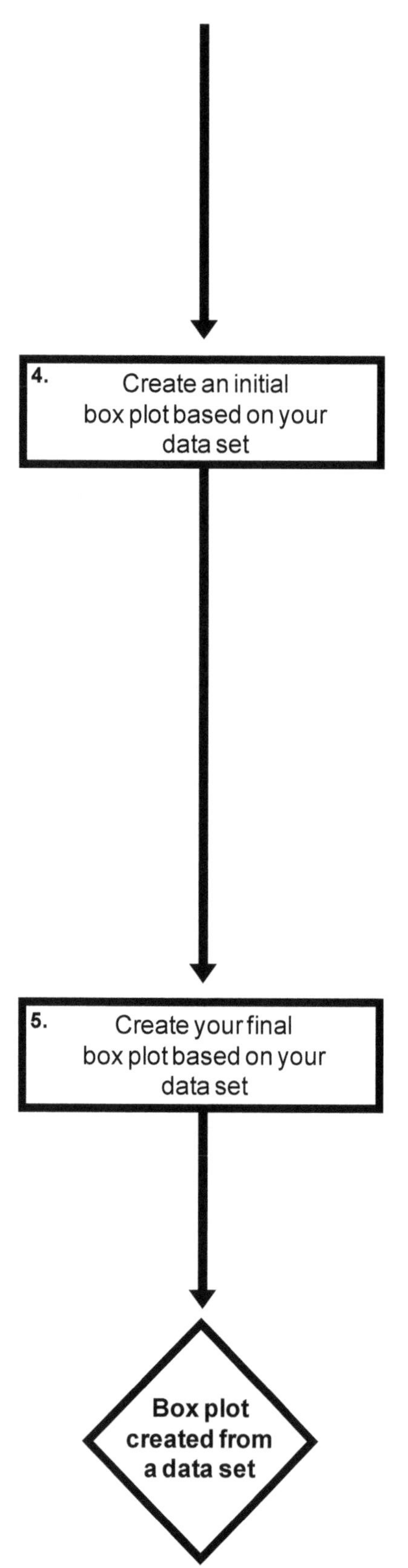
4.
Create an initial
box plot based on your
data set
5.
Create your final
box plot based on your
data set
Box plot
created from
a data set

Once you have checked that your data have been correctly imported into R, you are ready to create your initial box plot. You will use this initial box plot to check that your graph looks like you need it to look. To do this, enter the following command into R:

```
boxplot(clutch_size~species, data=boxplot_
data)
```

This is CODE BLOCK 48 in the document R_CODE_BASIC_STATS_WORKBOOK.DOC. This command creates a box plot from the column called `clutch_size` in the R object called `boxplot_data` created in step 2 of this exercise. On this graph, each box represents one category in the `species` column of this R object. As a result, this box plot will show the range of values for clutch size for the two species within the data set: blue tits (BT) and great tits (GT). At this stage, you can review your graph to ensure that it is showing exactly the information you wish it to show. If it doesn't, you may need to process your data set before you make your box plot from it. For example, you may need to subset it to remove data from additional species, check it for outliers that may represent errors, or process it to remove missing values before you re-make your initial box plot.

Once you have checked your initial box plot and you are happy that it shows the required information, you are ready to create the final version of it. To do this, enter the following command into R:

```
boxplot(clutch_size~species, data=
boxplot_data, xlab="Species", ylab="Clutch
Size", main="Recorded Clutch Size in Tit
Species", names=c("Blue Tit","Great Tit"))
```

This is CODE BLOCK 49 in the document R_CODE_BASIC_STATS_WORKBOOK.DOC. This code adds four new arguments to the `boxplot` command. These provide a label for the X axis (specified by the `xlab` argument), a label for the Y axis (specified by the `ylab` argument), a title for the box plot itself (specified by the `main` argument), and labels for the categories on the X axis (specified by the `names` argument).

The box plot produced by working through this example should look like this:

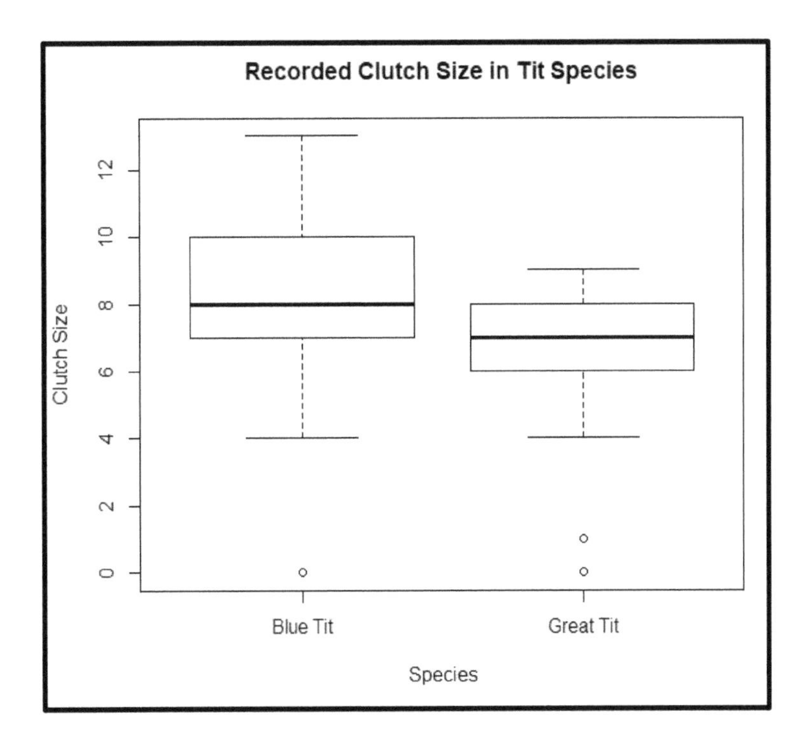

Once you have created a box plot, you can export it from R so that you can include it in a manuscript or presentation. If you are using RGUI, you can do this by clicking on the R GRAPHICS window containing your box plot to select it, before clicking on FILE on the main menu bar and selecting SAVE AS. This will allow you to save it in a variety of different formats. If you are using RStudio, you can export your graph by clicking on the EXPORT button at the top of the window displaying your box plot and selecting SAVE AS IMAGE.

When including a box plot in a manuscript, it is important that you write an appropriate figure legend for it. This legend should provide all the information required for the reader to interpret the contents of the graph. For the above box plot, an appropriate legend would be:

Figure 1: *Clutch sizes recorded for blue tits (left) and great tits (right) in nest boxes used to study the breeding success of hole-nesting birds. Central bar: Median clutch size; Box: Interquartile range; Whiskers: Full range of the data, excluding outliers; Circles: Outlying values.*

The `boxplot` command used to create the box plots in this exercise can be modified using a range of additional arguments to create a graph with the exact characteristics you wish it to have. The additional arguments that biologists most commonly use when creating box plots are provided in the table below, while the full list of additional arguments that can be used with the `boxplot` command can be found at *www.rdocumentation.org/packages/graphics/versions/3.6.1/topics/boxplot.*

Additional Argument	**How To Use It**
`col`	This argument allows you to specify the fill colour for the boxes on a box plot. The options for this argument can be the name of the desired colour, or a list of colours if you wish each box plotted on your graph to have a different fill colour. For example, for a two-group box plot, if you include the argument `col=c("blue","red")` it will create a box plot where the box representing the first series is coloured blue while the box representing the second series is coloured red. Alternatively, you can use a hexadecimal code to specify your desired colour. For example, including the argument `col=("#ff3300")` would create a box plot with red boxes for all the groups plotted on it. You can find a full list of hexadecimal codes for different colours at *www.color-hex.com.*
`names`	This argument allows you to specify the names that will be displayed on the X axis below the box for each group. The option that you will most often use for this argument is the name of a column in the R object on which the graph is based. For example, the argument `names=species` would create a box plot where the labels displayed below the bars along the X axis are obtained from a column in your data set called `species`. Alternatively, you can provide the names you wish to use for each data series as a list. For example, you could use the argument `names=c("Blue Tit","Great Tit")` to use these as the names for a box plot with two groups plotted on it.
`main`	This argument allows you to specify the text that will appear as the title of your box plot. The option for this argument is the text you wish to use for your title. For example, including the argument `main="Recorded Clutch Size in Tit Species"` will create a box plot with this text as the title above it. **NOTE**: This additional argument is optional, and in many cases you will include this information in your figure legend rather than on the graph itself.
`xlab`	This argument allows you to specify the label for the X axis of your box plot. The option for this argument is the text you wish to use as the label for it. For example, including the argument `xlab="Species"` will create a box plot with this label on its X axis.
`ylab`	This argument allows you to specify the label for the Y axis of your box plot. The option for this argument is the text you wish to use as the label for it. For example, including the argument `ylab="Clutch Size"` will create a box plot with this label on its Y axis.
`ylim`	This argument allows you to specify the minimum and maximum values for the Y axis of your box plot. The options for this argument can be any pair of numbers. For example, including the argument `ylim=c(0,15)` would create a box plot with an Y axis that has values ranging from 0 to 15, while including the argument `ylim=c(4,25)` would create a box plot with a Y axis that has values ranging from 4 to 25.

Additional Argument	How To Use It
`range`	This argument allows you to specify how far the 'whiskers' extend out from the either end of the central box on your box plot. The options for this argument are either 0 or a positive numeric value. If a value of 0 is used, the whiskers will cover the full range of the data. If the value is a positive number, the whiskers are limited to only covering the interquartile range multiplied by this number. For example, using the argument `range=0.5` will create a box plot where the whiskers for each box will extent 0.5 times the interquartile range above and below the main box.
`width`	This argument allows you to specify the width for the boxes on your box plot. The most commonly used option for this argument is a list of the relative widths for each box. For example, using the argument `width=c(100,50)` for a graph with two boxes on it would set the width of the second box to half the width of the first.

For the next part of this exercise, you will customise your `boxplot` command in a number of different ways. Firstly, you will add the `col` additional argument to set the fill colours used for each data series on the graph. To do this, add the argument `col=c("blue","red")` to the code provided in step 5 of the above flow diagram. The modified code should look like this (required modifications are highlighted in **bold**):

```
boxplot(clutch_size~species, data=boxplot_data,
xlab="Species", ylab="Clutch Size", main="Recorded Clutch
Size in Tit Species", names=c("Blue Tit","Great Tit"),
col=c("blue","red"))
```

You can modify this code either by editing it in the R CONSOLE window of RGUI or through the SCRIPT EDITOR window of RStudio (depending on which interface you are using). If you are entering commands directly into the R CONSOLE window, you can use the UP arrow on your keyboard to bring commands you have previously run during the same session back on to the command line of this window, and then use the LEFT and RIGHT arrows to scroll through and edit them. In this case, use the UP arrow to bring the previous version of the `boxplot` command back onto the command line and edit it so that it looks like the one above. Once you have finished modifying your `boxplot` command, you can run it by pressing the ENTER key on your keyboard. If you are using RStudio, you can copy and paste the original `boxplot` command in the SCRIPT EDITOR window before editing the new version to include the required modifications. Once you have done this, you can select the modified version of the command and click on the RUN button to run it in the R CONSOLE window. This modified code should produce a new box plot that looks like this:

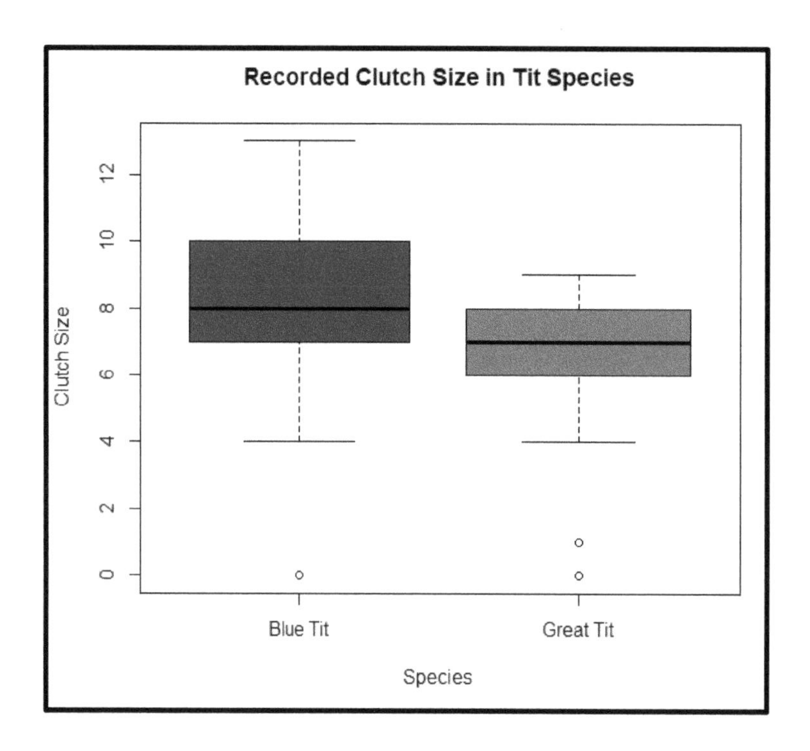

Once you have learned more about how to use the `boxplot` command by customising it, you can test your ability to use it correctly by typing the following command into R:

```
boxplot(clutch_size~species, data=boxplot_data,
xlab="Species", ylab="Clutch Size", main="Recorded Clutch
Size in Tit Species", names=c("Blue Tit","Great Tit"),
col=c("grey","white"), ylim=c(0,15))
```

This command will create a box plot with a grey box for the data on clutch size in blue tits and a white box for the data from great tits. A new additional argument has been added to this command to set the range of values that are displayed on the Y axis (using the `ylim` additional argument). When you run this version of the `boxplot` command, it should produce a new box plot that looks like the image at the top of the next page.

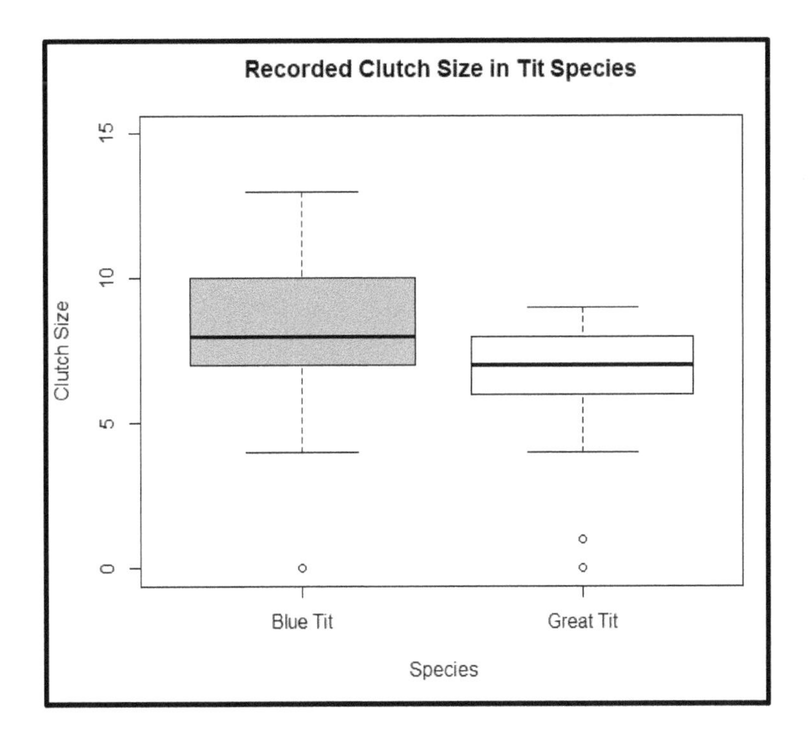

EXERCISE 2.5: HOW TO MAKE A SCATTER PLOT FROM BIOLOGICAL DATA:

So far, the graphs you have created in this chapter have provided summaries of the data in a data set. However, sometimes you will wish to plot the individual data points. Doing this allows you to assess whether there is a relationship between two variables and, if there is, what form this relationship takes. If these are continuous or ordinal variables, this can be done using a scatter plot. These are also known as X-Y plots and they show the distribution of data with respect to two axes, each of which represents a single variable. In order to be able to create a scatter plot in R, your data need to be arranged in a spreadsheet or table where each row contains data for a single record in your data set. In this table, there also needs to be separate columns containing values for the two continuous or ordinal variables. These are the variables that will be plotted on the X and the Y axes of your scatter plot. For this exercise, you will start by creating a scatter plot showing the relationship between land elevation and clutch size in blue tits from data collected from nest boxes in central Scotland. To do this, work through the flow diagram that starts at the top of the next page.

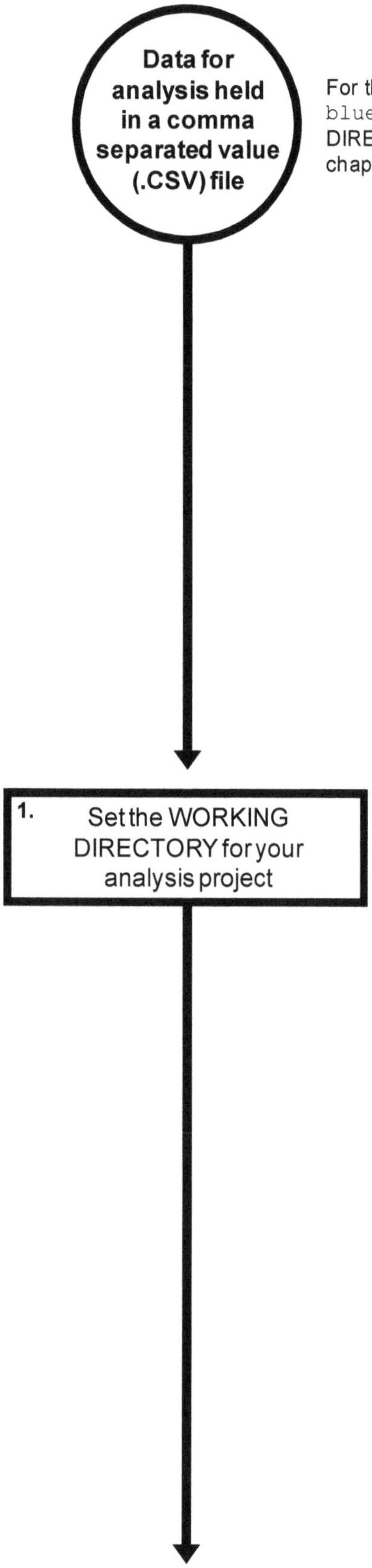

Data for analysis held in a comma separated value (.CSV) file

For this exercise, the data set you will use is stored in a file called `blue_tit_breeding_data.csv` that is located in the WORKING DIRECTORY folder you created during the introduction to this chapter.

1. Set the WORKING DIRECTORY for your analysis project

Before you start any analysis in R, you first need to set the WORKING DIRECTORY. To do this, enter the text `setwd("` and then type the address of your WORKING DIRECTORY, using slashes (/) as the folder separators, before entering a second quotation mark followed by a closing bracket, like this `")`. For example, if your WORKING DIRECTORY has the address C:\STATS_FOR_BIOLOGISTS_ONE, your `setwd` command should look like this:

```
setwd("C:/STATS_FOR_BIOLOGISTS_ONE")
```

If you are using RGUI, enter your `setwd` command in the R CONSOLE window (remembering to use the address of your own WORKING DIRECTORY folder in it) and then press the ENTER key on your keyboard. If you are using RStudio, enter your `setwd` command into the SCRIPT EDITOR window. To run it, select it and then click on the RUN button at the top of this window. You will enter all the remaining commands for this exercise in a similar manner, depending on the user interface you are using.

To check that your WORKING DIRECTORY has been set properly, enter the command `getwd()` and carefully check that the address it returns is the same as the one for the STATS_FOR_BIOLOGISTS_ONE folder you created at the start of this chapter.

Before you move on to step 2, make sure that all the data you wish to use in your analysis project are located in this WORKING DIRECTORY folder. In this case, this is a file called `blue_tit_breeding_data.csv`. **NOTE**: If the data you are going to import into R in step 2 are not located in the WORKING DIRECTORY you set in this step, the import code provided in the next step will not work.

2. Load your data into R using the `read.table` command

The `read.table` command provides the easiest way to import data held in a .CSV file into R so you can analyse it. To do this for the data set being used in this example, enter the following command into R:

```
blue_tits <- read.table(file="blue_tit_
breeding_data.csv", sep=",", as.is=FALSE,
header=TRUE)
```

This code has to be entered exactly as it is written here or it will not work. If you wish to use the copy-and-paste approach for entering this command, copy the text directly below CODE BLOCK 50 in the document R_CODE_BASIC_STATS_WORKBOOK.DOC and paste it into R.

This command will create a new object in R called `blue_tits` which will contain the data from the specified .CSV file. To load a different .CSV file into R, all you need to do is change the file name after the `file` argument to the name of the one you wish to import. In addition, you can use whatever name you wish for the R object which will be created by this command. To do this, simply replace `blue_tits` at the start of the first line of the above code with the name you wish to use for it. **NOTE:** If your .CSV data set uses a semicolon as the decimal separator, you would need to replace the `sep=","` argument with `sep=";"`.

3. Check the data have been loaded into R correctly by checking the names of the columns and by viewing it

Whenever you import any data into R, you need to check that they have loaded correctly. First, you need to check that all the required columns are present in the R object you just created. To do this, enter the following command into R:

```
names(blue_tits)
```

This is CODE BLOCK 51 in the document R_CODE_BASIC_STATS_WORKBOOK.DOC. This command will return the names used for each column in the R object created in step 2. For this example, the names should be `box_number`, `elevation`, `el_cat`, `species`, `clutch_size`, `no_chicks_hatched` and `no_chicks_fledged`.

Next, you should view the contents of the whole table using the `View` command. This is done by entering the following code into R:

```
View(blue_tits)
```

This is CODE BLOCK 52 in the document R_CODE_BASIC_STATS_WORKBOOK.DOC. This command will open a DATA VIEWER window where you can examine your data set and check that the correct data have been loaded into R.

4. Create an initial scatter plot based on your data set

Once you have checked that your data have been correctly imported into R, you are ready to create your initial scatter plot. You will use this to check that your graph looks like you need it to look. To do this, enter the following command into R:

```
plot(blue_tits$elevation, blue_tits$clutch_size)
```

This is CODE BLOCK 53 in the document R_CODE_BASIC_STATS_WORKBOOK.DOC. This `plot` command create a scatter plot based on the values in the columns called `elevation` and `clutch_size` for each data point in the R object called `blue_tits` created in step 2 of this exercise. At this stage, you can review your graph to ensure that it is showing exactly the information you wish it to show. If it doesn't, you may need to process your data set before you make your final scatter plot from it. For example, you may need to subset it to remove data from additional species, check it for outliers that may represent errors, or process it to remove missing values before you re-make your initial scatter plot.

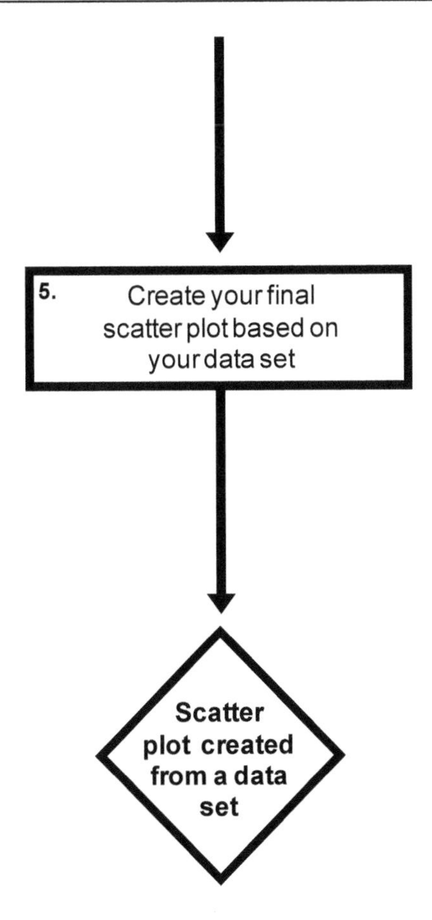

Once you have checked your initial scatter graph and you are happy that it shows the required information, you are ready to create the final version. This will add new settings to add labels to it. To do this, enter the following command into R:

```
plot(blue_tits$elevation, blue_tits$clutch_
  size, xlab="Elevation (M)", ylab="Clutch
  Size", main="Clutch Size vs Elevation in
             Blue Tits")
```

This is CODE BLOCK 54 in the document R_CODE_BASIC_STATS_WORKBOOK.DOC. This code adds three new arguments to the `plot` command from step 4. These provide a label for the X axis (specified by the `xlab` argument), a label for the Y axis (specified by the `ylab` argument) and a title for the scatter plot (specified by the `main` argument).

At the end of this exercise, you should have a scatter plot that looks like this:

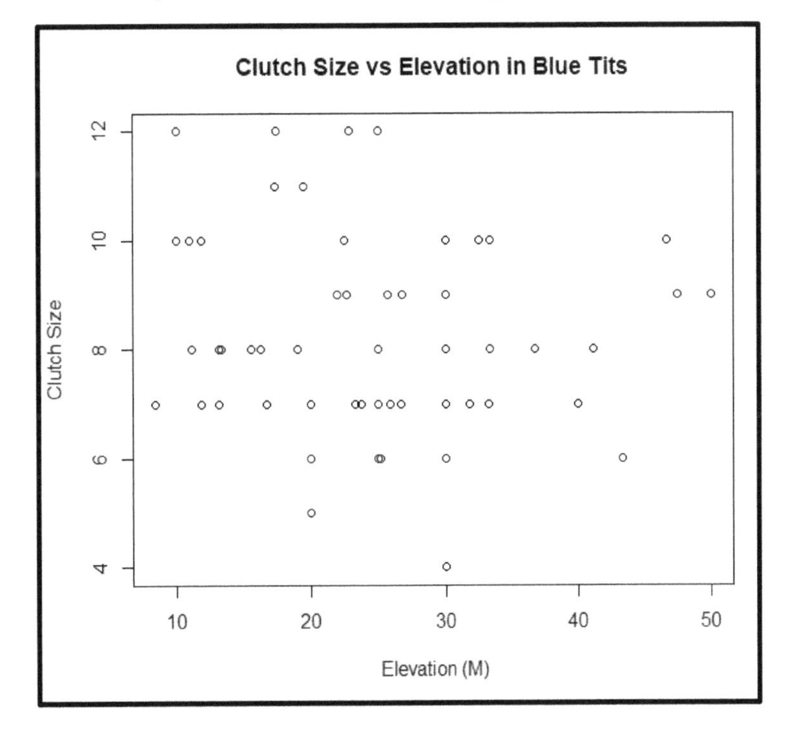

Once you have created a scatter plot, you can export it from R so that you can include it in a manuscript or presentation. If you are using RGUI, you can do this by clicking on the R GRAPHICS window containing your scatter graph to select it, before clicking on FILE on the main menu bar and selecting SAVE AS. This will allow you to save it in a variety of different formats. If you are using RStudio, you can export your graph by clicking on the EXPORT button at the top of the window displaying your scatter plot and selecting SAVE AS IMAGE.

When including a scatter plot in a manuscript, it is important that you write an appropriate figure legend for it. This legend should provide all the information required for the reader to interpret the contents of the graph. For the above graph, an appropriate legend would be:

Figure 1: *The relationship between the land elevation (in metres) where a nest box is sited and the clutch size recorded for blue tits in a sample of nest boxes.*

The `plot` command used to create the scatter plots in this exercise can be modified using a range of additional arguments to create a graph with the exact characteristics you wish it to have. The additional arguments that biologists most commonly use when creating scatter plots are provided in the table below, while a full list of the additional arguments that can be used with the `plot` command can be found at *www.rdocumentation.org/packages/graphics/versions/3.6.1/topics/plot*).

Additional Argument	How To Use It
`col`	This argument allows you to specify the colour used for the points on your scatter plot. The option for this argument can be the name of the desired colour. For example, including the argument `col="blue"` would create a graph with blue symbols on it. Alternatively, you can use a hexadecimal code to specify your desired colour. For example, including the argument `col="#ff3300"` would create a graph with red symbols on it. You can find a full list of hexadecimal codes for different colours at *www.color-hex.com*.
`pch`	This argument allows you to specify the shape and fill of the symbols shown on your scatter plot. For example, including the argument `pch=1` would create a graph with open circular symbols, while including the argument `pch=19` would create a graph with closed circular symbols. A list of all other possible values and the symbols they refer to can found at *www.sthda.com/english/wiki/r-plot-pch-symbols-the-different-point-shapes-available-in-r*.

Additional Argument	How To Use It
cex	This argument allows you to specify the size of the symbols shown on your scatter plot. The option for this argument can be any positive numeric value. For example, including the argument `cex=1` would create a graph with a symbol size of 1, while including the argument `cex=2` would create a graph with larger symbols on it. To get sizes in between, you can use decimal fractions (e.g. `cex=1.5`).
xlim	This argument allows you to specify the exact minimum and maximum values for the X axis of your scatter plot. The options for this argument can be any pair of numbers. For example, including the argument `xlim=c(0,70)` would create a graph with an X axis that has values ranging from 0 to 70, while including the argument `xlim=c(20,90)` would create a graph with an X axis that has values ranging from 20 to 90.
ylim	This argument allows you to specify the exact minimum and maximum values for the Y axis of your scatter plot. The options for this argument can be any pair of numbers. For example, including the argument `ylim=c(0,15)` would create a graph with a Y axis that has values ranging from 0 to 15, while including the argument `ylim=c(4,25)` would create a graph with a Y axis that has values ranging from 4 to 25.
main	This argument allows you to specify the text that will appear as a title at the top of the scatter plot. The option for this argument is the text you wish to use for your title. For example, including the argument `main="Clutch Size vs Elevation in Blue Tits"` will create a graph with this text as the title above it. **NOTE:** This additional argument is optional, and in many cases you will include this information in your figure legend rather than on the graph itself.
xlab	This argument allows you to specify the label for the X axis of your scatter plot. The option for this argument is the text you wish to use as the label for it. For example, including the argument `xlab="Elevation (M)"` will create a graph with this label on its X axis.
ylab	This argument allows you to specify the label for the Y axis of your scatter plot. The option for this argument is the text you wish to use as the label for it. For example, including the argument `ylab="Clutch Size"` will create a graph with this label on its Y axis.

For the next part of this exercise, you will customise your `plot` command in a number of different ways. Firstly, you will alter the colour used for your data points using the `col` additional argument. To do this, add the argument `col="blue"` to the `plot` command in step 5 of the above flow diagram.

The modified code for doing this should look like this (required modifications are highlighted in **bold**):

```
plot(blue_tits$elevation, blue_tits$clutch_size,
xlab="Elevation (M)", ylab="Clutch Size", main="Clutch Size
vs Elevation in Blue Tits", col="blue")
```

You can modify this code either by editing it in the R CONSOLE window of RGUI or through the SCRIPT EDITOR window of RStudio (depending on which interface you are using). If you are entering commands directly into the R CONSOLE window, you can use the UP arrow on your keyboard to bring commands you have previously run during the same session back on to the command line of this window, and then use the LEFT and RIGHT arrows to scroll through and edit them. In this case, use the UP arrow to bring the previous version of the `plot` command back onto the command line and edit it so that it looks like the one above. Once you have finished modifying your `plot` command, you can run it by pressing the ENTER key on your keyboard. If you are using RStudio, you can copy and paste the original `plot` command in the SCRIPT EDITOR window before editing the new version to include the required modifications. Once you have done this, you can select the modified version of the command and click on the RUN button to run it in the R CONSOLE window. This modified command should produce a new scatter plot that looks like this:

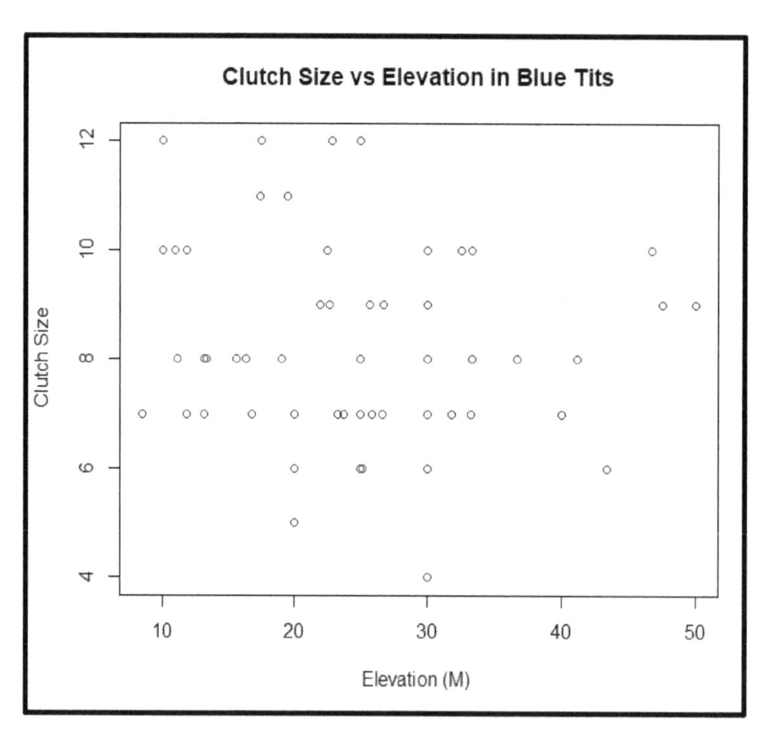

Next, you will alter the symbols used for your data points. This is done using the `pch` argument to change the symbol used to represent your data and the `cex` argument to change its size. To do this, add the arguments `pch=19` (the code for a filled circle) and `cex=1.5` to the `plot` command in step 5 of the above flow diagram. The modified `plot` command should look like the code provided at the top of the top of the next page.

```
plot(blue_tits$elevation, blue_tits$clutch_size,
xlab="Elevation (M)", ylab="Clutch Size", main="Clutch Size
vs Elevation in Blue Tits", col="blue", pch=19, cex=1.5)
```

Once you have finished modifying this command, you can run it. This should produce a new scatter plot that looks like this:

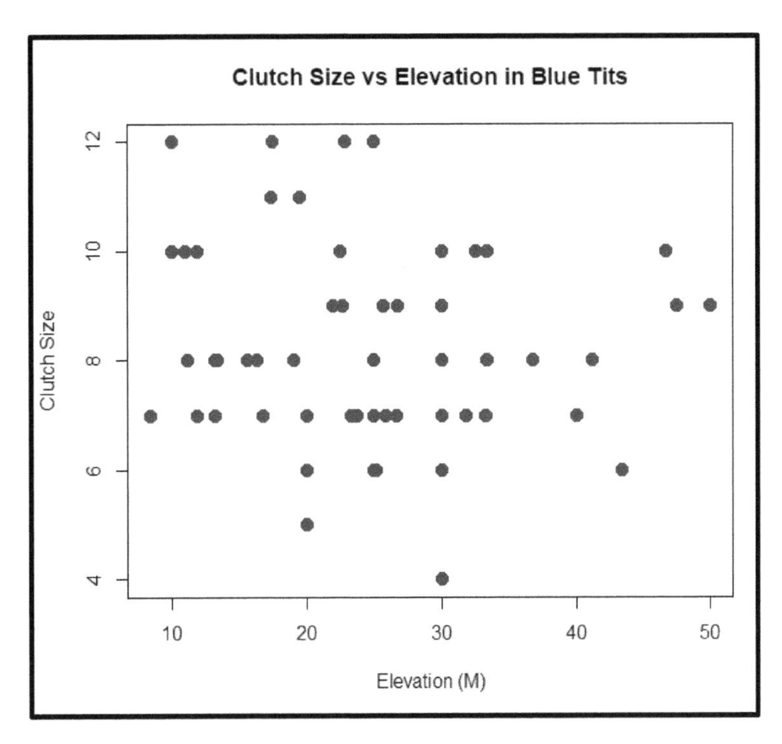

You can also customise a scatter plot by adding a line of best fit to it. This is done by adding a new command called abline to the code used to make the graph. To add a line of best fit to the above scatter plot, enter the following line of code into R immediately after you have run the plot command:

```
abline(lm(clutch_size~elevation, data=blue_tits),
                col="blue", lwd=2)
```

In this abline command, the lm argument defines the model used to fit the line (in this case, a simple linear regression model – see chapter 7 for more information), while the data argument sets the data set that will be used, the col argument sets the colour for the line, and the lwd argument sets the width of the line. When you run this command, it will add a line of best fit to the above scatter plot, and it should look like the image at the top of the next page.

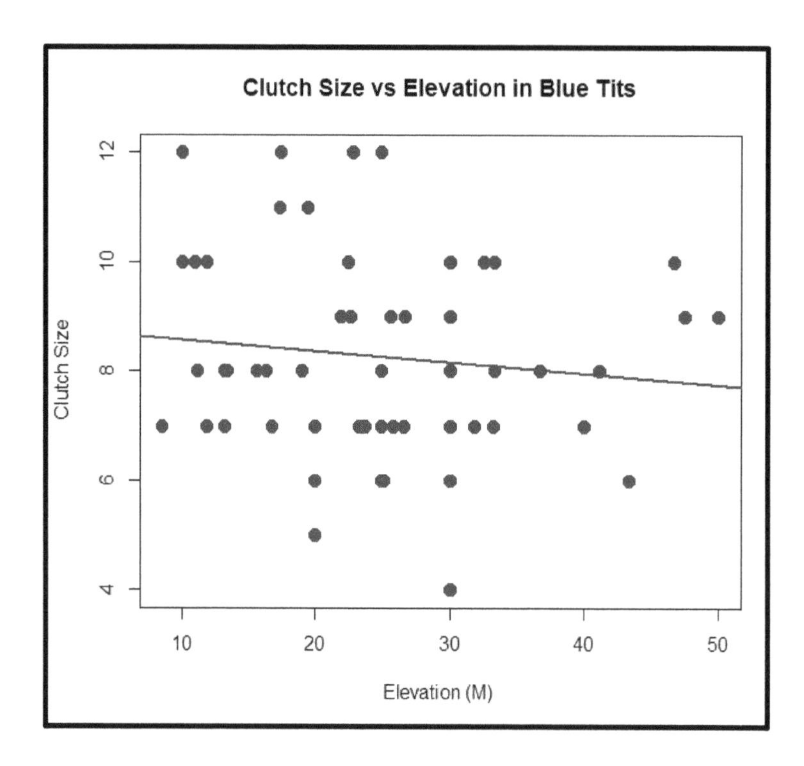

If you add a line of best fit to a scatter plot that you are creating for a manuscript, you will need to add information to the figure legend so that your reader knows what this line represents. For the above scatter plot, an appropriate legend would be:

Figure 1: *The relationship between the land elevation (in metres) and the clutch size recorded for blue tits from a sample of nest boxes. Solid Line: The line of best fit for the relationship between these variables.*

You can also customise your scatter plot so that it displays two series of data on the same graph. However, before you can do this, you would need import a second data set into R. To practice doing this, you can use the commands provided below to import information about breeding success from a second species, the great tit, collected from the same sample of nest boxes that the information on blue tits came from (**NOTE**: The .CSV file containing this data set must be located in the WORKING DIRECTORY you are using for your analysis).

```
great_tits <- read.table(file="great_tit_breeding_
   data.csv", sep=",", as.is=FALSE, header=TRUE)
View(great_tits)
```

Once you have imported your new data set, you need to re-plot your graph using the first data set by re-running the `plot` command for it. However, when you do this, you will use the `col="black"` argument to set the colour of the symbols to black rather than blue (as you did before). You will also need to drop the `cex` arguments and modify the title of the resulting scatter plot to reflect the fact that it will eventually have data from two species of tits plotted on it. This is done by changing the text after the `main` argument. The modified code to allow you to do this should look like this:

```
plot(blue_tits$elevation,blue_tits$clutch_size,
xlab="Elevation (M)", ylab="Clutch Size", main="Clutch Size
vs Elevation in Tit Species", pch=19, col="black")
```

You also need to re-run the `abline` command. However, you will also add the argument `col="black"` to this command to set the colour for the line to black. The modified version of this code should look like this:

```
abline(lm(clutch_size~elevation, data=blue_tits),
col="black")
```

At this stage, you can use the `points` command to plot a second data series on the same graph. For this example, use the following version of the `points` command to add the information on the relationships between land elevation and clutch size in great tits to your scatter plot:

```
points(great_tits$elevation, great_tits$clutch_size,
col="black", pch=1)
```

By using a different option for the `pch` argument (in this case, `pch=1` – the code for an open circle – rather than `pch=19` – the code for a closed circle) and/or the `col` argument, you can ensure that you can tell the difference between the two data sets on the resulting graph. Once you have added your second data series to your graph, you can also add a line of best fit for this new data series using this version of the `abline` command:

```
abline(lm(clutch_size~elevation, data=great_tits),
col="black", lty="dotted")
```

By including the additional `lty` argument in the `abline` command, you can set your second line to be dotted (rather than solid), making it easy to tell it apart from the line of best fit for the first data set. After you have finished adding your second series and its associated line of best fit, your scatter plot should look like this:

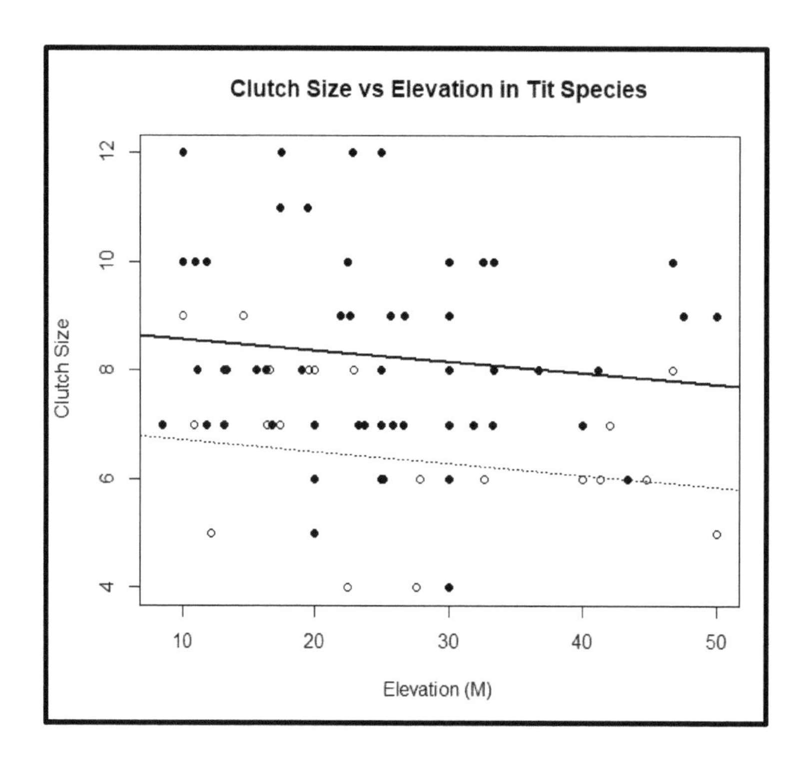

When creating a figure legend for graphs with more than one series on it, you need to make it clear which data are represented by which symbols. For this scatter plot, an appropriate legend would be:

Figure 1: *The relationship between the land elevation (in metres) and clutch size for two tit species based on data collected from a sample of nest boxes. Closed circles: Blue tits (solid line represents the line of best fit); Open circles: Great tits (dotted line represents the line of best fit).*

Once you have learned more about how to use the `plot` command by customising it, you can test your ability to use it correctly by entering the modified `plot` and `abline` commands provided at the top of the next page into R.

```
plot(blue_tits$elevation,blue_tits$no_chicks_fledged,
xlab="Elevation (M)", ylab="No Fledglings", main="Number of
Fledglings vs Elevation in Blue Tits", col="black", pch=15,
cex=1.5)
abline(lm(no_chicks_fledged~elevation, data=blue_tits),
col="black", lwd=2)
```

These commands will create a scatter plot showing the relationship between the number of blue tit fledglings (rather than clutch size) and the land elevation where each nest box was sited. This will use a black colour for the symbols (set by the `col="black"` argument), solid square symbols (set by the `pch=15` argument), and a line of best fit (specified by the `abline` command). When you run these commands, they should produce a new scatter plot that looks like this:

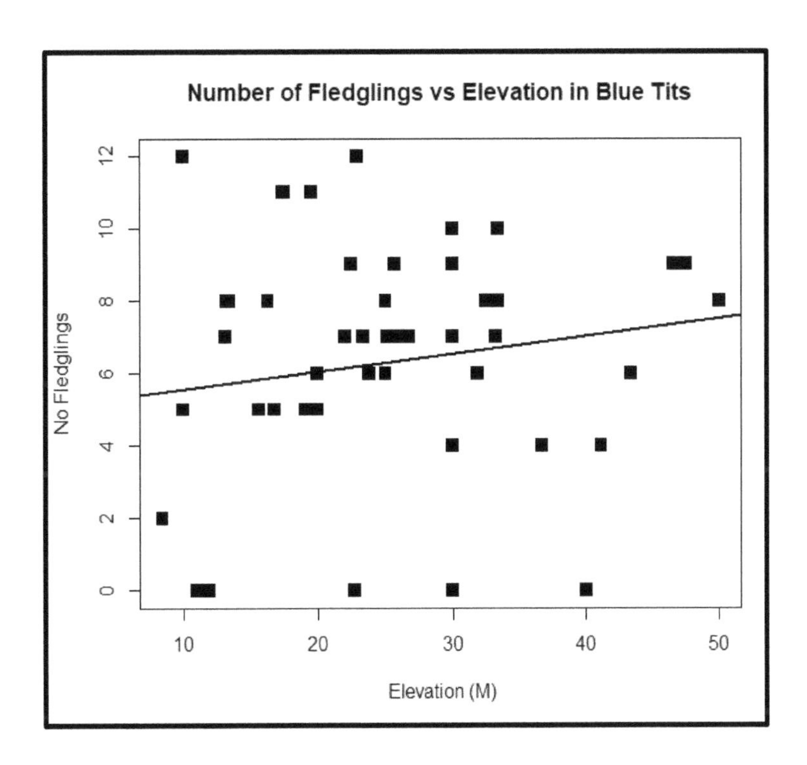

--- Chapter Five ---

Assessing And Transforming The Distribution Of Biological Data Using R

Once you have examined your data and created graphs from them, you will most likely want to start doing statistical analysis to find out whether they support any hypotheses you wish to investigate. However, before you can do this, you need to explore your data in more detail and check that they conform to the requirements of any specific statistical tests you wish to apply to them. For example, for parametric statistical tests, such as t-tests or ANOVAs, one of the most important requirements is that your data must have a normal distribution (see Figure 1). If your data are not normally distributed, then it is inappropriate to apply such tests to them. This means that in almost all instances, the first check you will run on a data set is a normality test.

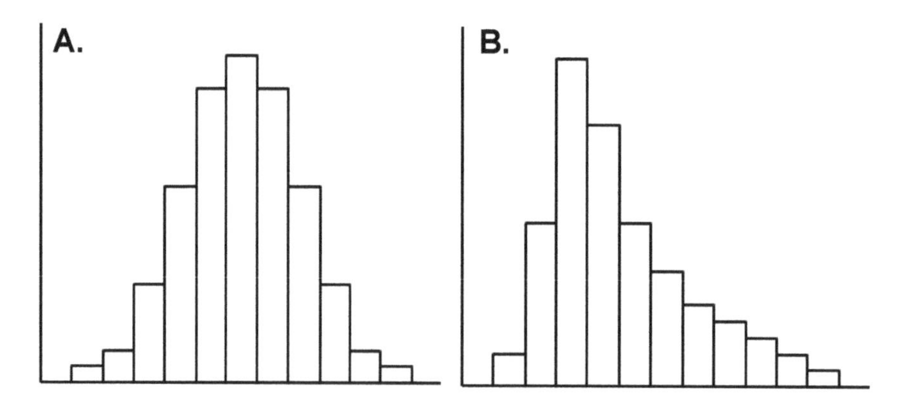

Figure 1: *Data with a normal distribution have a frequency distribution histogram that is more-or-less symmetrical around their mean value, like Histogram A, while data with a non-normal distribution have a non-symmetrical frequency distribution histogram, like Histogram B.*

If, when you test the normality of your data, you find that they are not normal, it may be possible to use a mathematical transformation to normalise their distribution. If such a transformation does result in data with a normal distribution, you can then apply a parametric test to them. If you cannot find a transformation that successfully normalises the distribution of your data, then you can use a non-parametric test, such as a Mann-Whitney U test or a Kruskal-Wallis test instead. While non-parametric tests are not as powerful as their parametric equivalents, they do allow you to analyse data with almost any type of distribution.

In this chapter, you will learn how to assess the distribution of a data set using tests for normality. In addition, you will learn how to apply mathematical transformations to a data set to see if this results in a normal distribution. In each case, you will learn about the structure that your data need to have to conduct a normality test or transformation. In addition, you will learn about all the steps you need to do to carry it out, starting with a data set held in a spreadsheet or table, and finishing with how to write up what you have done.

If you have not already done so, before you start the exercises in this chapter, you first need to create a WORKING DIRECTORY folder on your computer and load the necessary data into it (**<u>NOTE</u>:** If you have already created this folder and downloaded data for a previous chapter in this workbook, you do not need to do this again). To do this on a computer with a Windows operating system, open Windows Explorer and navigate to the location where you would like to create the folder (such as your C:\ drive or your DOCUMENTS folder). Next, right click anywhere in this location and select NEW> FOLDER. Now call this folder STATS_FOR_BIOLOGISTS_ONE by typing this into the folder name section to replace what it is currently called (which will most likely be NEW FOLDER). To create a WORKING DIRECTORY folder on a computer running a Mac operating system, open Finder and navigate to the location where you would like to create the folder (such as your DOCUMENTS folder or your DESKTOP). Next, click on FILE> NEW FOLDER, and then type the name STATS_FOR_BIOLOGISTS_ONE before pressing the ENTER key on your keyboard.

Once you have created your WORKING DIRECTORY folder, you are ready to download the data sets you will use for the exercises in this workbook from *www.gisinecology.com/stats-for-biologists-1*. After you have downloaded the compressed folder containing the required data by following the instructions provided on that page, you need to extract all the data files

from it and copy them into the folder called STATS_FOR_BIOLOGISTS_ONE that you have just created.

Next, you need to check that the required data have been extracted to the correct folder. If you are using a computer with a Windows operating system, you can use Windows Explorer to open your newly created WORKING DIRECTORY folder and examine its contents. If all the files from the compressed folder are present in it (there should be a total of 21 of them), you can click on the folder icon at the left hand end of the ADDRESS BAR at the top of the WINDOWS EXPLORER window to reveal its full address. Write this address down as you will need it to set this folder as your WORKING DIRECTORY during the exercises provided in this workbook (see pages 12 and 13 for details of how to modify folder addresses so they will be recognised by R).

If you are using a computer with a Mac operating system, you can use Finder to open your newly created WORKING DIRECTORY folder and examine its contents. If all the required data files are present in it (there should be a total of 21 of them), select this folder in Finder and then press the CMD and I keys on your keyboard at the same time. This will open the GET INFO window where you will find its address (which is also called the pathway). Write this address down somewhere as you will need it to set this folder as your WORKING DIRECTORY during the exercises provided in this workbook (see pages 12 and 13 for details of how to modify folder addresses so they will be recognised by R).

After you have loaded the required data into your WORKING DIRECTORY folder, you can open RGUI or RStudio, depending on which option you wish to use (see Chapter 2 for more details). Once you have opened your preferred R user interface, you need to create a file called CHAPTER_FIVE_EXERCISES where you will save the results of your analyses from your R CONSOLE window as you work through this chapter. To do this using RGUI, click on the FILE menu and select SAVE WORKSPACE. To do this in RStudio, click on SESSION and select SAVE WORKSPACE AS. In both cases, save it as a WORKSPACE file with the name CHAPTER_FIVE_EXERCISES.RDATA in your WORKING DIRECTORY folder (this will be the one called STATS_FOR_BIOLOGISTS_ONE that you have just created). If you are using RStudio, you will also want to save the contents of your SCRIPT EDITOR window (where you will enter and edit the R code you will use to carry out specific commands). To do this, click on the FILE menu and select SAVE AS.

Save your file as an R SCRIPT file with the name CHAPTER_FIVE_EXERCISES.R in your WORKING DIRECTORY folder. As you work through the exercises in this chapter, remember to regularly save the contents of your R CONSOLE window (which will contain the R objects you have created up to that point) to your WORKSPACE file and, if you are using RStudio, the contents of your SCRIPT EDITOR window to your R SCRIPT file.

Finally, you need to remove any data that are currently held in R's temporary memory. To do this, enter the following command into R (if you wish to copy and paste this command, the required code is directly below the text CODE BLOCK 1 in the document called R_ CODE_BASIC_STATS_WORKBOOK.DOC that is included in the compressed folder you just downloaded):

```
rm(list=ls())
```

If you are using RGUI, you can simply type or paste this code after the command prompt at the bottom of the R CONSOLE window (it looks like this: >) and then press the ENTER key on your keyboard to run it. If you are using RStudio, you can type or paste this command into the SCRIPT EDITOR window (the upper left hand window). To run this command, select it and then click on the RUN button at the top of this window. This will run it in the R CONSOLE window (the lower left hand one in the main RStudio user interface). You are now ready to start the exercises in this chapter.

EXERCISE 3.1: HOW TO ASSESS WHETHER A BIOLOGICAL DATA SET HAS A NORMAL DISTRBUTION USING R:

One of the main factors that will determine which statistical tests you can apply to a specific data set is whether or not the data being tested have a normal distribution. While you can get an idea of whether the data in a particular data set is likely to have a non-normal distribution by plotting a frequency distribution histogram (see Exercise 2.1) such assessments are subjective. As a result, it is important that you also apply an objective test to assess whether there really is a significant difference from normality.

There are three tests that are commonly used by biologists for doing this. These are the Shapiro-Wilk test, the Anderson-Darling test and the Kolmogorov-Smirnov test. Of these, the Shapiro-Wilk test is generally considered the most powerful, but it is also the one that is most sensitive to the proportion of tied or non-unique values within the data being examined. The Kolmogorov-Smirnov test is generally considered to be the least powerful, and is no longer recommended by many statisticians as a normality test for biological data. It does, however, allow you compare your data not just to a normal distribution, but to any distribution you wish to specify, and as a result is still retains enough usefulness for biologists to be included here.

Regardless of the test you wish to use, you need to have your data arranged in a spreadsheet or table where each row contains data from a single record in your data set. In this table, there also needs to be a column which contains values for the continuous variables you wish to test for normality. For this exercise, you will test whether or not two variables in a data set from a small sample of human males have distributions that differ significantly from normal. You will start by using a Shapiro-Wilk test to assess whether data on body mass from these men have a normal distribution. To do this, work through the flow diagram that starts on the next page.

For this example, the data set you will use is stored in a file called `human_data.csv` that is located in the WORKING DIRECTORY folder you created during the introduction to this chapter.

1. Set the WORKING DIRECTORY for your analysis project

Before you start any analysis in R, you first need to set the WORKING DIRECTORY. To do this, enter the text `setwd("` and then type the address of your WORKING DIRECTORY, using slashes (/) as the folder separators, before entering a second quotation mark followed by a closing bracket, like this `")`. For example, if your WORKING DIRECTORY has the address C:\STATS_FOR_BIOLOGISTS_ONE, your `setwd` command should look like this:

```
setwd("C:/STATS_FOR_BIOLOGISTS_ONE")
```

If you are using RGUI, enter your `setwd` command in the R CONSOLE window (remembering to use the address of your own WORKING DIRECTORY folder in it) and then press the ENTER key on your keyboard. If you are using RStudio, enter your `setwd` command into the SCRIPT EDITOR window. To run it, select it and then click on the RUN button at the top of this window. You will enter all the remaining commands for this exercise in a similar manner, depending on the user interface you are using.

To check that your WORKING DIRECTORY has been set properly, enter the command `getwd()` and carefully check that the address it returns is the same as the one for the STATS_FOR_BIOLOGISTS_ONE folder you created at the start of this chapter.

Before you move on to step 2, make sure that all the data you wish to use in your analysis project are located in this WORKING DIRECTORY folder. In this case, this is a file called `human_data.csv`. **NOTE:** If the data you are going to import into R in step 2 are not located in the WORKING DIRECTORY you set in this step, the import code provided in the next step will not work.

2. Load your data into R using the `read.table` command

The `read.table` command provides the easiest way to import data held in a .CSV file into R so you can analyse it. To do this, you will use the following command:

```
human_data <- read.table(file="human_
data.csv", sep=",", as.is=FALSE,
header=TRUE)
```

This code has to be entered exactly as it is written here or it will not work. If you wish to use the copy-and-paste approach for entering this command, copy the text directly below CODE BLOCK 55 in the document R_CODE_BASIC_STATS_WORKBOOK.DOC and paste it into R.

This command will create a new object in R called `human_data` which will contain the data from the specified .CSV file. To load a different .CSV file into R, all you need to do is change the file name in the `file` argument to the name of the one you wish to import. In addition, you can use whatever name you wish for the object which will be created by this command. To do this, simply replace `human_data` at the start of the first line of the above code with the name you wish to use for it. **NOTE:** If your .CSV data set uses a semicolon as the decimal separator, you would need to replace the code `sep=","` with the code `sep=";"`.

3. Check the data have loaded into R correctly by checking the names of the columns and by viewing it

Whenever you import any data into R, you need to check that they have loaded correctly. First, you need to check that all the required columns are present in the R object you just created. To do this, enter the following command into R:

```
names(human_data)
```

This is CODE BLOCK 56 in the document R_CODE_BASIC_STATS_WORKBOOK.DOC. This command will return the names used for each column in the R object created in step 2. For this example, the names should be `id, mass, height` and `no_offspring`.

Next, you should view the contents of the whole table using the `View` command. This is done by entering the following code into R:

```
View(human_data)
```

This is CODE BLOCK 57 in the document R_CODE_BASIC_STATS_WORKBOOK.DOC. This command will open a DATA VIEWER window where you can examine your data set and check that the correct data have been loaded into R.

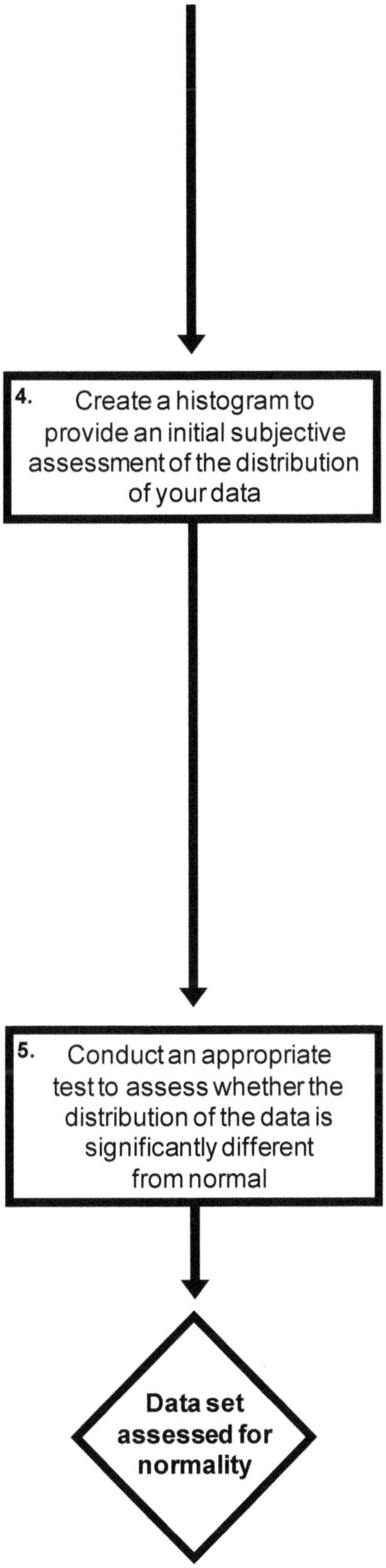

4. Create a histogram to provide an initial subjective assessment of the distribution of your data

Once your data have been successfully imported into R, you are ready to start assessing whether or not it has a distribution that differs significantly from normal. The first step in this process is to plot a frequency distribution histogram that you can examine to provide an initial subjective assessment of the distribution of your data. To do this, enter the following command into R:

```
hist(human_data$mass, nclass=7)
```

This is CODE BLOCK 58 in the document R_CODE_BASIC_STATS_WORKBOOK.DOC. This `hist` command creates a frequency distribution histogram from the data in the column called `mass` in the R object called `human_data` created in step 2 of this exercise. The term `nclass=7` in this command means that the data will be divided into approximately seven equal-sized classes to make this histogram (see Exercise 2.1 for more details). For your own data sets, you may wish to use a different value for the `nclass` argument.

Once you have created your frequency distribution histogram, you can examine it to see whether it looks like it may be normal, or if it is clearly non-normal (see figure 1). This histogram can also be used to assess whether your data violate the assumptions of any particular normality test (such as the proportion of data points with identical values). If, based on your histogram, you suspect that your data may violate the assumptions of a particular normality test, you can use the `View` command from step 3 to allow you to explore your data in more detail.

5. Conduct an appropriate test to assess whether the distribution of the data is significantly different from normal

Once you have conducted a subjective assessment of the distribution of your data, and you have ensured that they do not violate the requirements of the normality test you wish to use, you are ready to apply it to your data set. In this example, you will apply a Shapiro-Wilk test to assess whether or not the data on body mass from a sample of human males differs significantly from normal. To do this, enter the following command into R:

```
shapiro.test(human_data$mass)
```

This is CODE BLOCK 59 in the document R_CODE_BASIC_STATS_WORKBOOK.DOC. This code applies the Shapiro-Wilk test to data contained in the column called `mass` in the R object called `human_data` created in step 2 of this exercise.

Data set assessed for normality

At the end of the first part of this exercise, the frequency distribution histogram created in step 4 should look like this:

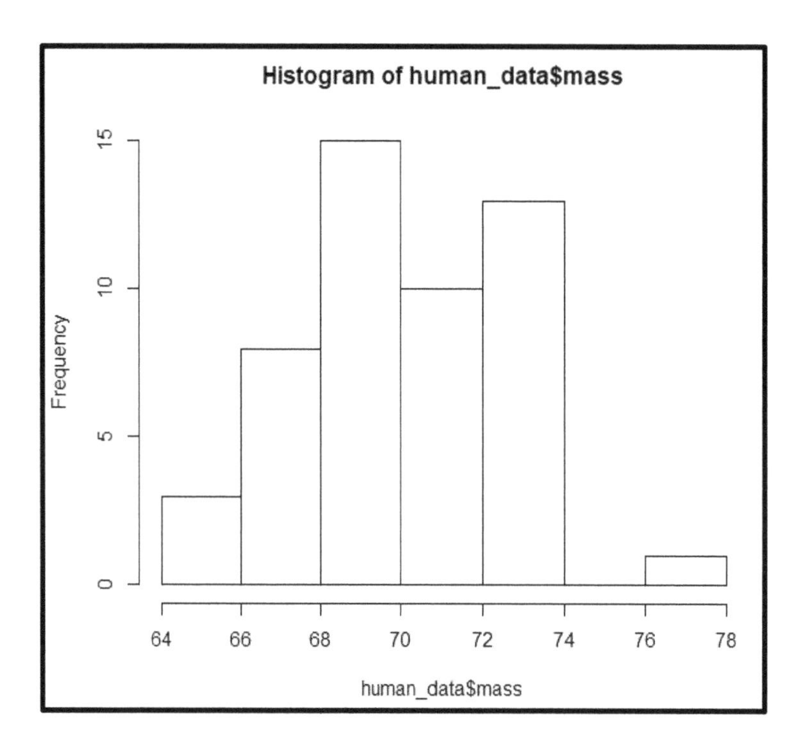

While the contents of your R CONSOLE window should look like this:

```
> setwd("C:/STATS_FOR_BIOLOGISTS_ONE")
> getwd()
[1] "C:/STATS_FOR_BIOLOGISTS_ONE"
> human_data <- read.table(file="human_data.csv", sep=",", header=TRUE)
> names(human_data)
[1] "id"            "mass"          "height"        "no_offspring"
> View(human_data)
> hist(human_data$mass,nclass=7)
> shapiro.test(human_data$mass)

        Shapiro-Wilk normality test

data:  human_data$mass
W = 0.97892, p-value = 0.5073

> |
```

If you examine the results of this Shapiro-Wilk test, you will see that the p-value for this test is 0.5073. This means there is no significant difference between the distribution of the body mass data from this sample of human males and a normal distribution. As a result, it would be appropriate to apply a parametric test, such as a t-test, to these data. To report the results of a Shapiro-Wilk test in a manuscript, you need to provide the value of the test statistic (W), the associated p-value (p), and the sample size (n). For the above example, you could report it as follows (with all values rounded to an appropriate number of figures – see Appendix III for details):

There was no significant difference between the distribution of the data on body mass from a sample of male humans and a normal distribution (Shapiro-Wilk Test: $W=0.98$; $p=0.507$; $n=50$).

When assessing the distribution of your data for normality using the above workflow, you can use any suitable normality test in step 5 as long as your data do not violate its assumptions. Information on the three normality tests most commonly used by biologists, along with the commands you can use to run them in R, are provided in the table below.

Normality Test	How To Conduct It In R
Shapiro-Wilk Test	To conduct a Shapiro-Wilk test, you can use the `shapiro.test` command. For this command, you need to define the column and the R object that contains the data for the continuous variable you wish to test for normality. For example, to run this test on the contents of a column called `mass` in an R object called `human_data`, the full command would be: `shapiro.test(human_data$mass)` If you need to refine your Shapiro-Wilk test, there are a number of additional arguments that can be used with this command. Details of these additional arguments can be found at *www.rdocumentation.org/ packages/stats/versions/3.6.1/topics/shapiro.test.*
Anderson-Darling Test	To conduct an Anderson-Darling test, you can use the `ad.test` command. This command is contained in the `goftest` package, and you will need to install this package and library before you can use it (see page 114 for more details). To use this command, you need to define the column and the R object that contains the data for the continuous variable you wish to test for normality. For example, to run this test on the contents of a column called `mass` in an R object called `human_data`, the full command would be: `ad.test(human_data$mass)` If you need to refine your Anderson-Darling test, there are a number of additional arguments that can be used with this command. Details of these additional arguments can be found at *www.rdocumentation.org/packages/goftest/versions/1.2-2/topics/ad.test.*

Normality Test	How To Conduct It In R
Kolmogorov-Smirnov (K.S.) Test	To conduct a Kolmogorov-Smirnov test, you can use the `ks.test` command. To use this command, you need to define the column and the R object that contains the data for the continuous variable you wish to test for normality. For example, to test whether contents of a column called `mass` in an R object called `human_data` differs significantly from a normal distribution (set by the argument `pnorm`), the full command would be: `ks.test(human_data$mass, pnorm)` If you need to refine your Kolmogorov-Smirnov test, there are a number of additional arguments that can be used with this command. Details of these additional arguments can be found at *www.rdocumentation.org/packages/stats/versions/3.6.1/ topics/ks.test.*

For the next part of this exercise, you will customise the above approach for conducting normality tests in a number of different ways. Firstly, you will run the same Shapiro-Wilk test on a different column of data in the same `human_data` R object. This column is called `no_offspring` and it contains data on the number of children that each male fathered in their lifetime. To run the second version of this test, you will first need to modify the code used in step 4 to create a frequency distribution histogram based on a different column. The modified code should look like this (required modifications are highlighted in **bold**):

```
hist(human_data$no_offspring, nclass=7)
```

You can modify this code either by editing it in the R CONSOLE window of RGUI or through the SCRIPT EDITOR window of RStudio (depending on which interface you are using). If you are entering commands directly into the R CONSOLE window, you can use the UP arrow on your keyboard to bring commands you have previously run during the same session back on to the command line of this window, and then use the LEFT and RIGHT arrows to scroll through and edit them. In this case, use the UP arrow to bring the previous version of the `hist` command back onto the command line and edit it so that it looks like the one above. Once you have finished modifying your `hist` command, you can run it by pressing the ENTER key on your keyboard. If you are using RStudio, you can copy and paste the original `hist` command in the SCRIPT EDITOR window before editing the new version to include the required modifications. Once you have done this, you can select the modified version of the command and click on the RUN button to run it in the R CONSOLE window.

When you run this modified version of the `hist` command, the histogram it creates should look like this (**NOTE:** Even though you set the `nclass` argument to equal seven, if you examine the resulting histogram, you will see that it only has six bars on it; this is okay, and is a result of R rounding down the number of classes to produce a frequency histogram with a more appropriate distribution than would be produced if it had seven classes in it – see Exercise 2.1 for more details):

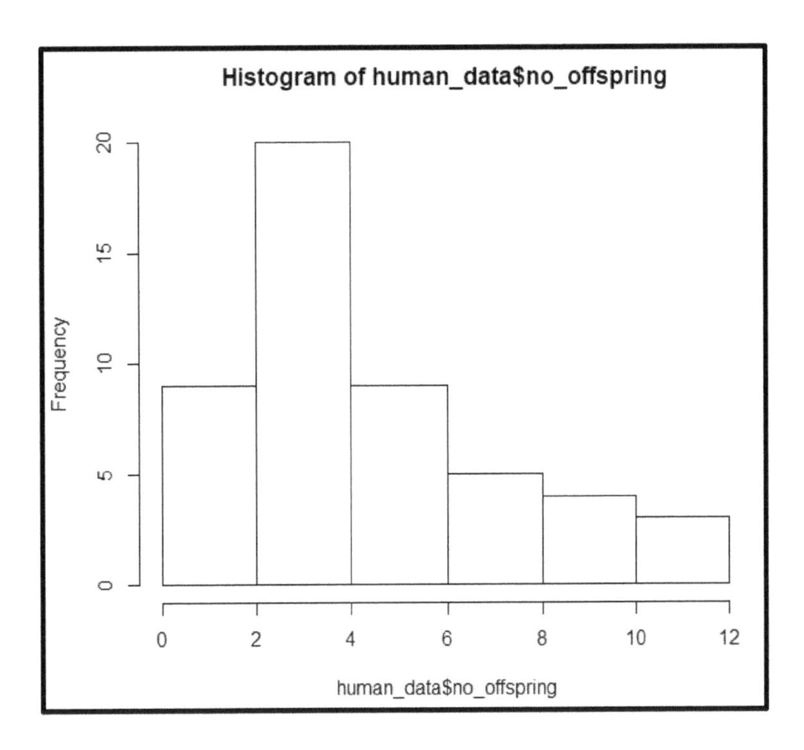

Once you have created the frequency distribution histogram based on the `no_offspring` column, you can then run a Shapiro-Wilk test on these data. To do this, you will need to modify the column referenced in the R code used in step 5. The modified code should look like this:

```
shapiro.test(human_data$no_offspring)
```

Once this command has been run, the contents of your R CONSOLE window should look like the image at the top of the next page.

```
> hist(human_data$no_offspring,nclass=7)
> shapiro.test(human_data$no_offspring)

        Shapiro-Wilk normality test

data:  human_data$no_offspring
W = 0.89588, p-value = 0.0003523

> |
```

If you examine the results of this Shapiro-Wilk test, you will see that the p-value for this test is 0.0003523. This means that there is a significant difference between the distribution of the number of children fathered by each male in this data set and a normal distribution. As a result, it would not be appropriate to apply a parametric test, such as a t-test, to these data. The results of this Shapiro-Wilk test can be reported as follows:

There was a significant difference between the distribution of the data on the number of offspring fathered by a sample of male humans and a normal distribution (Shapiro-Wilk Test: $W=0.90; p<0.001; n=50$).

If you wish to run a different test for normality, you can work through the same workflow and replace the R command in step 5 with the appropriate R command from the above table. To get experience in doing this, you can repeat the above normality test on the number of offspring data (held in the `no_offspring` column) in the data set `human_data` using an Anderson-Darling test rather than a Shapiro-Wilk test. However, before you can run an Anderson-Darling test, you will need to install the `goftest` package and then load the `goftest` library into R. To do this, first run the command:

```
install.packages("goftest")
```

Follow any instructions that appear, and once this package has been downloaded and installed, you can use the following command to load the appropriate library into R:

```
library(goftest)
```

After this library has been loaded into R, you can modify the code provided in step 5 of the above flow diagram so that it runs the Anderson-Darling test rather than the Shapiro-Wilk test. The modified code should look like the code at the top of the next page.

```
ad.test(human_data$no_offspring)
```

After you have run this command, the contents of your R CONSOLE window should look like this:

```
> install.packages("goftest")
Installing package into 'C:/Users/Computer/Documents/R/win-library/3.6'
(as 'lib' is unspecified)
trying URL 'https://cran.ma.imperial.ac.uk/bin/windows/contrib/3.6/goftest_1.2-2.zip'
Content type 'application/zip' length 75410 bytes (73 KB)
downloaded 73 KB

package 'goftest' successfully unpacked and MD5 sums checked

The downloaded binary packages are in
        C:\Users\Computer\AppData\Local\Temp\Rtmpm42Z2x\downloaded_packages
> library(goftest)
> ad.test(human_data$no_offspring)

        Anderson-Darling test of goodness-of-fit
        Null hypothesis: uniform distribution
        Parameters assumed to be fixed

data:  human_data$no_offspring
An = Inf, p-value = 1.2e-05

> |
```

As was the case when a Shapiro-Wilk test was applied to these data, this Anderson-Darling test found a significant difference between the data on the number of children fathered by this sample of human males and a normal distribution. To report the results of an Anderson-Darling test in a manuscript, you need to provide the value of the test statistic (An), the associated p-value (p), and the sample size (n). For the above example, you could report it as follows:

The distribution of the data on the number of offspring in a sample of male humans was found to differ significantly from a normal distribution (Anderson-Darling Test: An=Inf; p<0.001; n=50).

Once you have learned how to customise the workflow for conducting tests for normality on a data set, it, you can test your ability to use it correctly by applying a Kolmogorov-Smirnov test for normality to the same `no_offspring` data. To do this, enter the following modified version of the command from step 5 into the R:

```
ks.test(human_data$no_offspring, pnorm)
```

After you have run this command, the contents of your R CONSOLE window should look like the image at the top of the next page.

```
> ks.test(human_data$no_offspring, pnorm)

        One-sample Kolmogorov-Smirnov test

data:  human_data$no_offspring
D = 0.93725, p-value < 2.2e-16
alternative hypothesis: two-sided

Warning message:
In ks.test(human_data$no_offspring, pnorm) :
  ties should not be present for the Kolmogorov-Smirnov test
> |
```

As with the previous two tests on these same data, the Kolmogorov-Smirnov test also found that distribution of the number of offspring data from human males in this data set differed significantly from normal (p<2.2e-16, or 0.00000000000000022). To report the results of a Kolmogorov-Smirnov test, you need to provide the value of test statistic (D), the associated p-value (p), and the sample size (n). For the above example, you could report it as follows:

*The distribution of the data on the number of offspring fathered by a sample of male humans differed significantly from a normal distribution (*Kolmogorov-Smirnov *Test: D=0.94; p<0.001; n=50).*

EXERCISE 3.2: HOW TO NORMALISE BIOLOGICAL DATA USING A MATHEMATICAL TRANSFORMATION IN R:

If you find that a set of data has a distribution that differs significantly from normal, the first thing you should try is to see whether you can normalise it using a mathematical transformation. This is because parametric tests are generally more powerful than their non-parametric equivalents and if a transformation does allow you to normalise your data, it is often more appropriate to do this rather than simply applying a non-parametric test.

There are a range of different transformations that you can apply to see if they will normalise the distribution of your data, and the most appropriate one to use will depend on the exact distribution that your data have (see Figure 2). For this reason, it is important to plot a histogram of your data prior to selecting a transformation to apply to them. In addition, after you have done any transformations, you need to both view your data to check that the transformation has been conducted properly, and run a test for normality to ensure that the transformed data are now sufficiently normal to allow a parametric test to be applied to them.

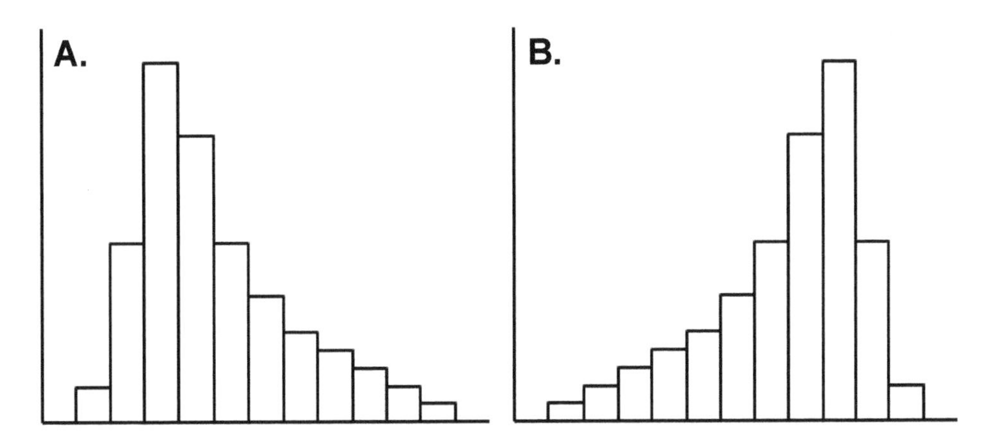

Figure 2: *The mathematical transformation you apply to a set of non-normal data will depend on the shape of its distribution. In particular, different transformations should be applied to data that have a right-skewed distribution, like Histogram A, and a left-skewed distribution, like Histogram B.*

In order to apply a mathematical transformation to a particular variable in a data set in R, you need to have your data arranged in a spreadsheet or table where each row contains data from a single record in your data set. In this table, there also needs to be a column which contains the values for the variable you wish to transform. In this exercise, you will apply a number of mathematical transformations to the set of data from a sample of human males used in Exercise 3.1. You will start by applying a log transformation to the data on the number of offspring fathered by each individual before testing whether this transformation has successfully normalised them. To do this, work through the flow diagram that starts at the top of the next page.

NOTE: This workflow assumes that you have already set your WORKING DIRECTORY, successfully loaded your data into R (in this instance, as an object called `human_data`), run a normality test on your data and found that the variable you are interested in analysing requires transformation – see Exercise 3.1 for details of how to do these steps.

For this example, the data set you will use is held in the R object called `human_data` created in Exercise 3.1 of this workbook.

1. Create a histogram to help you decide which mathematical transformation you should apply to your data

The first step in the process of applying a mathematical transformation to a data set in R is to plot a frequency distribution histogram. This will help you identify which mathematical transformations are appropriate to apply to your data. To create a frequency distribution histogram for the data in the column called `no_offspring` in the R object called `human_data`, enter the following command into R:

```
hist(human_data$no_offspring, nclass=7)
```

This code has to be entered exactly as it is written here or it will not work. If you wish to use the copy-and-paste approach for entering this command, copy the text directly below CODE BLOCK 60 in the document R_CODE_BASIC_STATS_WORKBOOK.DOC and paste it into R. Once you have created your histogram, you can examine it to assess which mathematical transformation will be most appropriate to apply to your data set (see the table on page 122).

2. Apply your selected transformation to the column of data you wish to transform

Once you have decided on the mathematical transformation you wish to use, you then need to apply it to your data. This is done by creating a new column in your R object and using the appropriate mathematical function to calculate the transformed values (see page 122 for details). For this example, you will apply a log transformation to the data in the column called `no_offspring` in the R object `human_data`. To do this, enter the following command into R:

```
human_data$log_no_offspring=log10(human_
        data$no_offspring)
```

This is CODE BLOCK 61 in the document R_CODE_BASIC_STATS_WORKBOOK.DOC. This command creates a new column called `log_no_offspring` in the R object called `human_data` and then fills it with values created by log-transforming the data in the `no_offspring` column. Once you have applied a mathematical transformation to your data, you should always check the results to ensure that they have been calculated correctly. To do this for the log-transformed number of offspring data, enter the following command into R:

```
View(human_data)
```

This is CODE BLOCK 62 in the document R_CODE_BASIC_STATS_WORKBOOK.DOC. This will allow you to examine your data set, complete with the newly added column containing the values for the transformed variable, to ensure that the transformation has been done correctly. You can do this by manually applying the required mathematical function to the first couple of values in the column of data you have transformed using a calculator and checking that you get the same result as R. If you do not, you need to repeat this step to ensure you have applied the required function correctly.

Once you are happy that your transformation has been done correctly, you can create a new histogram based on it so you can make a subjective assessment of how the distribution has changed. To do this for the data you transformed in this step, enter the following command into R:

```
hist(human_data$log_no_offspring, nclass=7)
```

This is CODE BLOCK 63 in the document R_CODE_BASIC_STATS_WORKBOOK.DOC.

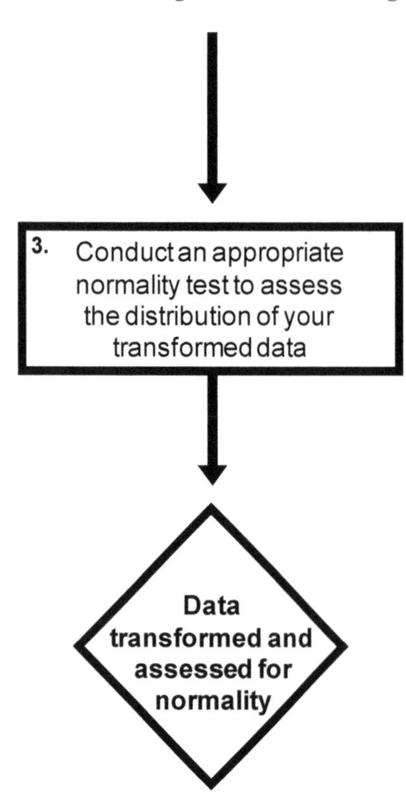
3. Conduct an appropriate normality test to assess the distribution of your transformed data
Data transformed and assessed for normality

Whenever you transform a column of data to try to normalise it, you should always run a normality test to ensure that the intended result has been achieved. In this example, you will apply a Shapiro-Wilk test to assess whether or not the transformed data differ significantly from normal. To do this, enter the following command into R:

```
shapiro.test(human_data$log_no_offspring)
```

This is CODE BLOCK 64 in the document R_CODE_BASIC_STATS_WORKBOOK.DOC. This code applies the Shapiro-Wilk test to the data contained in the column called `log_no_offspring` in the R object called `human_data`. This column was created in step 2 of this exercise.

At the end of the first part of this exercise, the frequency distribution histogram of your transformed variable (`log_no_offspring`) produced in step 2 should look like this:

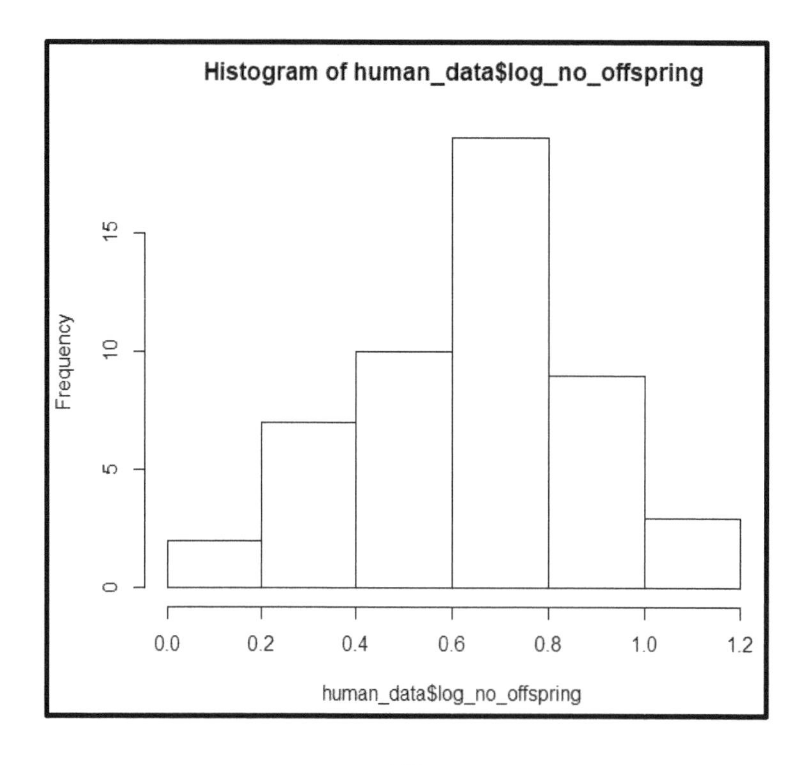
Histogram of human_data$log_no_offspring
Frequency
0
5
10
15
0.0
0.2
0.4
0.6
0.8
1.0
1.2
human_data$log_no_offspring

While the contents of your R CONSOLE window should look like this:

```
> hist(human_data$no_offspring,nclass=7)
> human_data$log_no_offspring=log10(human_data$no_offspring)
> View(human_data)
> hist(human_data$log_no_offspring,nclass=7)
> shapiro.test(human_data$log_no_offspring)

        Shapiro-Wilk normality test

data:  human_data$log_no_offspring
W = 0.9638, p-value = 0.1282

> |
```

If you examine the results of this Shapiro-Wilk test, you will see that there is no significant difference between the distribution of log-transformed number of offspring data and a normal distribution (p-value=0.1282). This confirms that the data transformation has been successful, and you could now apply a parametric test to these transformed data without violating any underlying assumptions of normality.

To report this transformation process, you would need to provide details of the mathematical transformation you applied and the results of the normality test you used to confirm that this transformation has successfully normalised your data, with all values rounded to an appropriate number of figures (see Appendix III for details). For the above example, you could report it as follows:

In order to normalise the data on the number of offspring fathered by different individuals in a sample of human males, a log transformation was applied to them. The transformed data did not differ significantly from a normal distribution (Shapiro-Wilk Test: W=0.96; p=0.128; n=50).

When transforming data using the above workflow, you can apply a range of different mathematical transformations in step 2. Your choice of transformation will primarily depend on the shape of the frequency distribution of your original untransformed data. Information on the mathematical data transformations most commonly used by biologists for right-skewed and left-skewed data (see Figure 2), along with the R commands required to apply them, are provided in the table on the next page.

Transformation	Command Required To Carry It Out In R (In All Cases, *X* Represents The Variable You Wish To Transform)

Log Transformation (for right-skewed data)	`log10(X)`
	For example, to transform the contents of a column called `no_offspring` in an R object called `human_data`, the full command would be:
	`human_data$log_no_offspring=log10(human_data$no_offspring)`
	NOTE: This transformation cannot be applied to any variable with values of 0 in it. To log transform such variables, you would need to use the formula `log10(X+1)`.

Square root Transformation (for right-skewed data)	`sqrt(X)`
	For example, to transform the contents of a column called `no_offspring` in an R object called `human_data`, the full command would be:
	`human_data$sqrt_no_offspring=sqrt(human_data$no_offspring)`

Arcsine Transformation (for right-skewed data)	`asin(X)`
	For example, to transform the contents of a column called `no_offspring` in an R object called `human_data`, the full command would be:
	`human_data$asin_no_offspring=asin(human_data$no_offspring)`
	NOTE: This transformation should only be applied to variables with values between 0 and 1. Typically, these will be proportional data.

Reciprocal Transformation (for right-skewed data)	`1/(X)`
	For example, to transform the contents of a column called `no_offspring` in an R object called `human_data`, the full command would be:
	`human_data$recip_no_offspring=1/(human_data$no_offspring)`

Squared Transformation (for left-skewed data)	`X^2`
	For example, to transform the contents of a column called `no_offspring` in an R object called `human_data`, the full command would be:
	`human_data$sq_no_offspring=(human_data$no_offspring^2)`

Exponential Transformation (for left-skewed data)	`exp(X)`
	For example, to transform the contents of a column called `no_offspring` in an R object called `human_data`, the full command would be:
	`human_data$exp_no_offspring=exp(human_data$no_offspring)`

For the next part of this exercise, you will customise the above approach for transforming a column of data in a number of different ways. Firstly, you will run the same log transformation on a different column of data in the `human_data` data set. This is the column of data called `height`. To do this, you will need to modify the commands used in step 2. The modified versions of these commands should look like this (required modifications are highlighted in **bold**):

```
human_data$log_height=log10(human_data$height)
View(human_data)
hist(human_data$log_height, nclass=7)
```

You can modify this code either by editing it in the R CONSOLE window of RGUI or through the SCRIPT EDITOR window of RStudio (depending on which interface you are using). If you are entering commands directly into the R CONSOLE window, you can use the UP arrow on your keyboard to bring commands you have previously run during the same session back on to the command line of this window, and then use the LEFT and RIGHT arrows to scroll through and edit them. In this case, use the UP arrow to bring the previous versions of these commands back onto the command line and edit them so that they look like the ones provided above. Once you have finished modifying each command, you can run it by pressing the ENTER key on your keyboard. If you are using RStudio, you can copy and paste the original commands in the SCRIPT EDITOR window before editing the new versions to include the required modifications. Once you have done this, you can select the modified versions of the commands and click on the RUN button to run them in the R CONSOLE window.

When you run this modified version of the `hist` command, the frequency distribution histogram created by it should look like the image at the top of the next page.

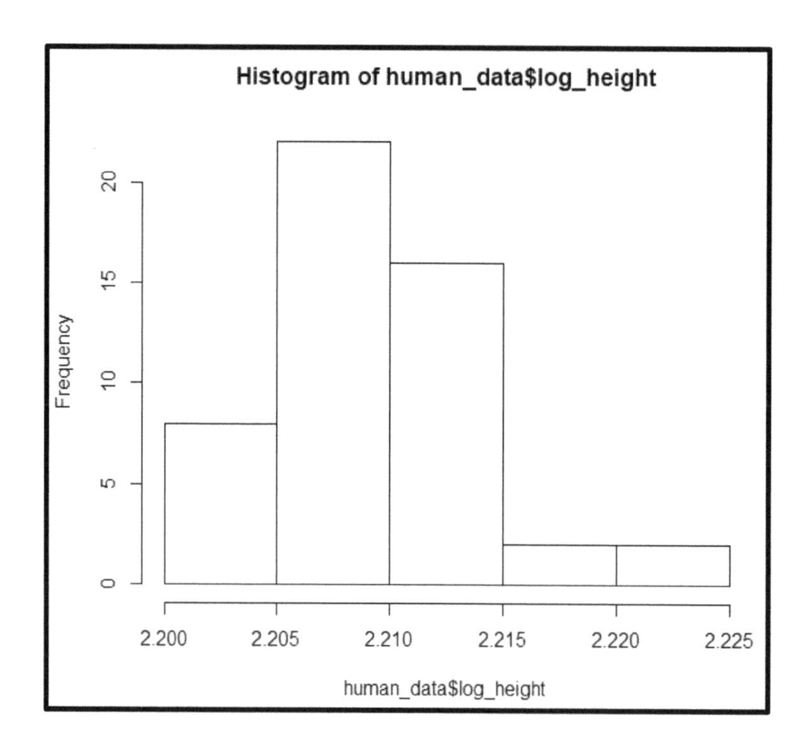

To run the Shapiro-Wilk test for normality on this log-transformed height data, you will need to modify the column referenced in the R code used in step 3. The modified code should look like this:

```
shapiro.test(human_data$log_height)
```

Once you have run this command, the contents of your R CONSOLE window should look like this:

```
> human_data$log_height=log10(human_data$height)
> View(human_data)
> hist(human_data$log_height,nclass=7)
> shapiro.test(human_data$log_height)

        Shapiro-Wilk normality test

data:  human_data$log_height
W = 0.91418, p-value = 0.001462

> |
```

If you examine the results of this Shapiro-Wilk test, you will see that there is still a significant difference between the distribution of the log-transformed height data and a normal distribution (p-value=0.001462). This means that your mathematical transformation has not achieved its aim of normalising the distribution of the height data. If, as in this case, you find that a particular transformation does not normalise your data, you can repeat the process using a different transformation to see if it is any more successful. To do this, replace the R command for the log transformation in step 2 with a different transformation command. For example, you could apply a reciprocal transformation to these height data by modifying the code from step 2 so that it looks like this:

```
human_data$recip_height=1/human_data$height
View(human_data)
hist(human_data$recip_height, nclass=7)
```

When you run these modified commands, the frequency distribution histogram created by it should look like this:

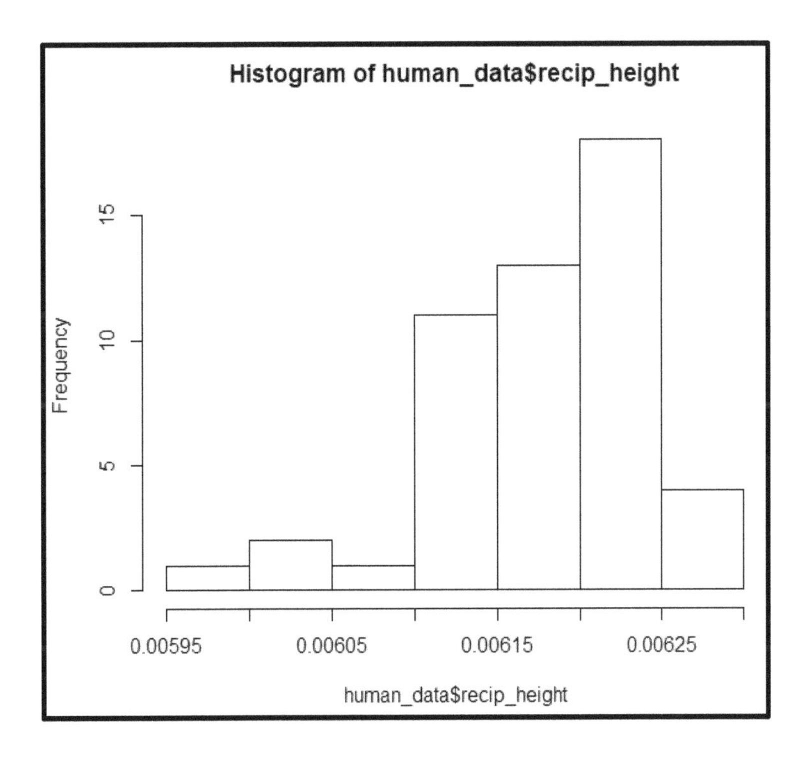

To run a Shapiro-Wilk test for normality on the reciprocally transformed height data, you will need to modify the column referenced in the R code used in step 3. The modified `shapiro.test` command to do this should look like the one at the top of the next page.

```
shapiro.test(human_data$recip_height)
```

After you have run this command, the contents of your R CONSOLE window should look like this:

```
> human_data$recip_height=1/human_data$height
> View(human_data)
> hist(human_data$recip_height,nclass=7)
> shapiro.test(human_data$recip_height)

        Shapiro-Wilk normality test

data:  human_data$recip_height
W = 0.9176, p-value = 0.001933

> |
```

If you examine the results of this Shapiro-Wilk test, you will see that there is still a significant difference between the distribution of the reciprocally transformed height data and a normal distribution (p-value=0.001933). This means that this second mathematical transformation still has not achieved its aim of normalising the distribution of these height data. Given the level of significance of this test, it is unlikely that any other transformation will result in the normalisation of these data. As a result, if you wish to analyse these data with a statistical test, you would have to use a non-parametric test (such as a Mann-Whitney U test) rather than a parametric one (such as a t-test).

NOTE: If you are unsure as to what the best transformation is for a given non-normal variable in a specific data set, this can be worked out empirically by applying a range of data transformations and seeing which one has a distribution that is closest to normal based on the p-value of a suitable normality test. For example, you could apply a log transformation, a square root transformation and a reciprocal transformation to a single variable and then select the transformation that gives the highest p-value in a normality test to help you determine which is most appropriate for your specific data set.

--- Chapter Six ---

Using Statistical Analysis To Compare Data From Different Groups In R

One of the main reasons for conducting statistical analyses on biological data is to investigate whether there are significant differences between two, or more, groups or samples. For example, you may wish to test whether males and females of the same species differ significantly in their body size, whether there is a significant difference in the growth of plants under control and experimental conditions or whether individuals that live under different environmental conditions have different body masses. There are a number of ways you can test for such differences, including testing for differences in mean or median values (known collectively as central values), testing for differences in the spread or variance of the data, and testing for differences in the number of records in each group in different categories.

Each of these comparisons requires the use of different statistical tests. For example, testing for differences in the mean of the data from two groups is often conducted with a t-test (see Exercise 4.1), while testing for differences in variance is usually conducted with an F-test or a Levene's test (see Exercise 4.3). Each of these tests have their own requirements and limitations, and these need to be checked to ensure your data conform to them before you apply a particular test. If you find that your data violate an important assumption of a particular test, then you need to select a different test. For example, a t-test requires that the data from the two groups you wish to compare have a normal distribution. If a normality test reveals that the data from one or both groups have a non-normal distribution, it would be inappropriate to conduct a t-test. Instead, you would have to use a non-parametric equivalent, such as a Mann-Whitney U test, which does not require the data to have a normal distribution.

In this chapter, you will learn how to compare data from different groups or samples using a variety of different statistical tests. For each type of test, you will start by importing your data into R, before checking the data against the requirements of a particular test, and running the most appropriate test given the characteristics of the data. Finally, at the end, you will be provided with advice on how to report the results of your analysis when writing up your findings.

If you have not already done so, before you start the exercises in this chapter, you first need to create a WORKING DIRECTORY folder on your computer and load the necessary data into it (**NOTE:** If you have already created this folder and downloaded data for a previous chapter in this workbook, you do not need to do this again). To do this on a computer with a Windows operating system, open Windows Explorer and navigate to the location where you would like to create the folder (such as your C:\ drive or your DOCUMENTS folder). Next, right click anywhere in this location and select NEW> FOLDER. Now call this folder STATS_FOR_BIOLOGISTS_ONE by typing this into the folder name section to replace what it is currently called (which will most likely be NEW FOLDER). To create a WORKING DIRECTORY folder on a computer running a Mac operating system, open Finder and navigate to the location where you would like to create the folder (such as your DOCUMENTS folder or your DESKTOP). Next, click on FILE> NEW FOLDER, and then type the name STATS_FOR_BIOLOGISTS_ONE before pressing the ENTER key on your keyboard.

Once you have created your WORKING DIRECTORY folder, you are ready to download the data sets you will use for the exercises in this workbook from *www.gisinecology.com/stats-for-biologists-1*. After you have downloaded the compressed folder containing the required data by following the instructions provided on that page, you need to extract all the data files from it and copy them into the folder called STATS_FOR_BIOLOGISTS_ONE that you have just created.

Next, you need to check that the required data have been extracted to the correct folder. If you are using a computer with a Windows operating system, you can use Windows Explorer to open your newly created WORKING DIRECTORY folder and examine its contents. If all the files from the compressed folder are present in it (there should be a total of 21 of them), you can click on the folder icon at the left hand end of the ADDRESS BAR at the top of the WINDOWS EXPLORER window to reveal its full address. Write this address down as you will need it to set this folder as your WORKING DIRECTORY during the exercises provided in this workbook (see pages 12 and 13 for details of how to modify folder addresses so they will be recognised by R).

If you are using a computer with a Mac operating system, you can use Finder to open your newly created WORKING DIRECTORY folder and examine its contents. If all the required data files are present in it (there should be a total of 21 of them), select this folder in Finder and then press the CMD and I keys on your keyboard at the same time. This will open the GET INFO window where you will find its address (which is also called the pathway). Write this address down somewhere as you will need it to set this folder as your WORKING DIRECTORY during the exercises provided in this workbook (see pages 12 and 13 for details of how to modify folder addresses so they will be recognised by R).

After you have loaded the required data into your WORKING DIRECTORY folder, you can open RGUI or RStudio, depending on which option you wish to use (see Chapter 2 for more details). Once you have opened your preferred R user interface, you need to create a file called CHAPTER_SIX_EXERCISES where you will save the results of your analyses from your R CONSOLE window as you work through this chapter. To do this using RGUI, click on the FILE menu and select SAVE WORKSPACE. To do this in RStudio, click on SESSION and select SAVE WORKSPACE AS. In both cases, save it as a WORKSPACE file with the name CHAPTER_SIX_EXERCISES.RDATA in your WORKING DIRECTORY folder (this will be the one called STATS_FOR_BIOLOGISTS_ONE that you have just created). If you are using RStudio, you will also want to save the contents of your SCRIPT EDITOR window (where you will enter and edit the R code you will use to carry out specific commands). To do this, click on the FILE menu and select SAVE AS. Save your file as an R SCRIPT file with the name CHAPTER_SIX_EXERCISES.R in your WORKING DIRECTORY folder. As you work through the exercises in this chapter, remember to regularly save the contents of your R CONSOLE window (which will contain the R objects you have created up to that point) to your WORKSPACE file and, if you are using RStudio, the contents of your SCRIPT EDITOR window to your R SCRIPT file.

Finally, you need to remove any data that are currently held in R's temporary memory. To do this, enter the following command into R (if you wish to copy and paste this command, the required code is directly below the text CODE BLOCK 1 in the document called R_ CODE_BASIC_STATS_WORKBOOK.DOC that is included in the compressed folder you just downloaded):

```
rm(list=ls())
```

If you are using RGUI, you can simply type or paste this code after the command prompt at the bottom of the R CONSOLE window (it looks like this: >) and then press the ENTER key on your keyboard to run it. If you are using RStudio, you can type or paste this command into the SCRIPT EDITOR window (the upper left hand window). To run this command, select it and then click on the RUN button at the top of this window. This will run it in the R CONSOLE window (the lower left hand one in the main RStudio user interface). You are now ready to start the exercises in this chapter.

EXERCISE 4.1: HOW TO TEST FOR A DIFFERENCE IN THE CENTRAL VALUES (MEANS/MEDIANS) OF TWO GROUPS:

One of the most common ways to test for a difference between data from two groups or samples is to compare their central values. In this context, the central values refer to either the mean values (if the data from both groups have a normal distribution) or the median values (if the distribution of the data from one or both groups differs significantly from normal). If the data from both groups have a normal distribution, you can use a t-test to investigate whether there is a significant difference in the mean value between them. If the data from one or both of your groups is non-normal, you can use a Mann-Whitney U test to investigate whether there is a significant difference between their median values

In order to be able to test for differences in the central values of two groups, you need to have your data in a spreadsheet or table with one row for each data point in your data set, regardless of which group it belongs to. In this table, you also need to have a column containing the continuous variable you wish to compare between your two groups and a second column which tells you which group each data point belongs to.

For this exercise, you will start by testing if there is significant difference between the body mass of male and female great tits. To do this, work through the following flow diagram:

For this example, the data set you will use is stored in a file called `great_tit_mass_data.csv` that is located in the WORKING DIRECTORY folder you created during the introduction to this chapter.

1. Set the WORKING DIRECTORY for your analysis project

Before you start any analysis in R, you first need to set the WORKING DIRECTORY. To do this, enter the text `setwd("` and then type the address of your WORKING DIRECTORY, using slashes (/) as the folder separators, before entering a second quotation mark followed by a closing bracket, like this `")`. For example, if your WORKING DIRECTORY has the address C:\STATS_FOR_BIOLOGISTS_ONE, your `setwd` command should look like this:

```
setwd("C:/STATS_FOR_BIOLOGISTS_ONE")
```

If you are using RGUI, enter your `setwd` command in the R CONSOLE window (remembering to use the address of your own WORKING DIRECTORY folder in it) and then press the ENTER key on your keyboard. If you are using RStudio, enter your `setwd` command into the SCRIPT EDITOR window. To run it, select it and then click on the RUN button at the top of this window. You will enter all the remaining commands for this exercise in a similar manner, depending on the user interface you are using.

To check that your WORKING DIRECTORY has been set properly, enter the command `getwd()` and carefully check that the address it returns is the same as the one for the STATS_FOR_BIOLOGISTS_ONE folder you created at the start of this chapter.

Before you move on to step 2, make sure that all the data you wish to use in your analysis project are located in this WORKING DIRECTORY folder. In this case, this is a file called `great_tit_mass_data.csv`. **NOTE**: If the data you are going to import into R in step 2 are not located in the WORKING DIRECTORY you set in this step, the import code provided in the next step will not work.

The `read.table` command provides the easiest way to load data held in a .CSV file into R so you can analyse it. To do this, you will use the following command:

```
great_tit_mass_data <- read.table(file=
  "great_tit_mass_data.csv", sep=",",
        as.is=FALSE,header=TRUE)
```

This code has to be entered exactly as it is written here or it will not work. If you wish to use the copy-and-paste approach for entering this command, copy the text directly below CODE BLOCK 65 in the document R_CODE_BASIC_STATS_WORKBOOK.DOC and paste it into R.

2. Load your data into R using the `read.table` command

This command will create a new object in R called `great_tit_mass_data` which will contain the data from the specified .CSV file. To load a different .CSV file into R, all you need to do is change the file name after the `file` argument to the name of the one you wish to import. In addition, you can use whatever name you wish for the object which will be created by this command. To do this, simply replace `great_tit_mass_data` at the start of the first line of the above code with the name you wish to use for it. **NOTE:** If your .CSV data set uses a semicolon as the decimal separator, you would need to replace the `sep=","` argument with `sep=";"`.

Whenever you import any data into R, you need to check that they have loaded correctly. First, you need to check that all the required columns are present in the R object you just created. To do this, enter the following command into R:

```
names(great_tit_mass_data)
```

3. Check the data have loaded into R correctly by checking the names of the columns and by viewing it

This is CODE BLOCK 66 in the document R_CODE_BASIC_STATS_WORKBOOK.DOC. This command will return the names used for each column in the R object created in step 2. For this example, the names should be `id`, `mass`, `wing_length`, `sex` and `age`.

Next, you should view the contents of the whole table using the `View` command. This is done by entering the following code into R:

```
View(great_tit_mass_data)
```

This is CODE BLOCK 67 in the document R_CODE_BASIC_STATS_WORKBOOK.DOC. This command will open a DATA VIEWER window where you can examine your data set and check that the correct data have been loaded into R.

4. Create a box plot to provide an initial subjective assessment of how the data from your two groups compare

Once your data have been successfully imported into R, you are ready to start assessing whether or not different groups in your data set differ significantly from each other. The first step in this process is to create a box plot of the data from the different groups. To do this, enter the following command into R:

```
boxplot(mass~sex, data=great_tit_mass_data)
```

This is CODE BLOCK 68 in the document R_CODE_BASIC_STATS_WORKBOOK.DOC. This command creates a box plot from the data in the column called `mass` in the R object called `great_tit_mass_data` created in step 2. This box plot will show the range of values for body masses for the two sexes within the data set based on the categories in the column called `sex`. This means it can be used to provide a subjective assessment of how the values for body mass within the two groups compare.

Before you can decide which test to apply to objectively compare the central values of your two groups, you first need to separate your data so you have different R objects containing the data from each group. This is so that you can run a separate normality test on the data from each one. If you find that the data from both groups have normal distributions, you can use a t-test to compare their mean values. If the data distribution of one or both groups is non-normal, the first thing you should do is try to normalise it using a mathematical transformation (see Exercise 3.2). Once you have applied your transformation you can then return to this workflow and repeat steps 4 and 5 for the transformed version of the variable. When you do this, if you find that the data for both groups are now normal, you can apply a t-test to them in step 6. If you cannot find a transformation that successfully normalises the data from both groups, you will need to use a Mann-Whitney U test to compare the medians of the two data sets.

In this example, you will first divide your great tit body mass data into two group based on the contents of the column called `sex`. To do this, enter the following code into R:

```
male_great_tit_data <- subset(great_tit_
        mass_data, sex=="male")
female_great_tit_data <- subset(great_
    tit_mass_data, sex=="female")
```

5. Assess whether or not each group of data in your data set has a normal distribution

This is CODE BLOCK 69 in the document R_CODE_BASIC_STATS_WORKBOOK.DOC.

Once you have created the subsets from your data based on the groupings you wish to compare, you can test each subset using an appropriate normality test (see Exercise 3.1). For the data being used in this example, you will use the Shapiro-Wilk test. To apply this test to the male great tit data, enter the following command into R:

```
shapiro.test(male_great_tit_data$mass)
```

This is CODE BLOCK 70 in the document R_CODE_BASIC_STATS_WORKBOOK.DOC. When you do this, you will find that the distribution of body mass data for male great tits is not significantly different from normal (p-value=0.5607).

This test is then repeated for the second subset of data. To do this for the female great tit data, enter the following command into R:

```
shapiro.test(female_great_tit_data$mass)
```

This is CODE BLOCK 71 in the document R_CODE_BASIC_STATS_WORKBOOK.DOC. When you do this, you will find that the distribution of body mass data for female great tits is also not significantly different from normal (p-value=0.4676).

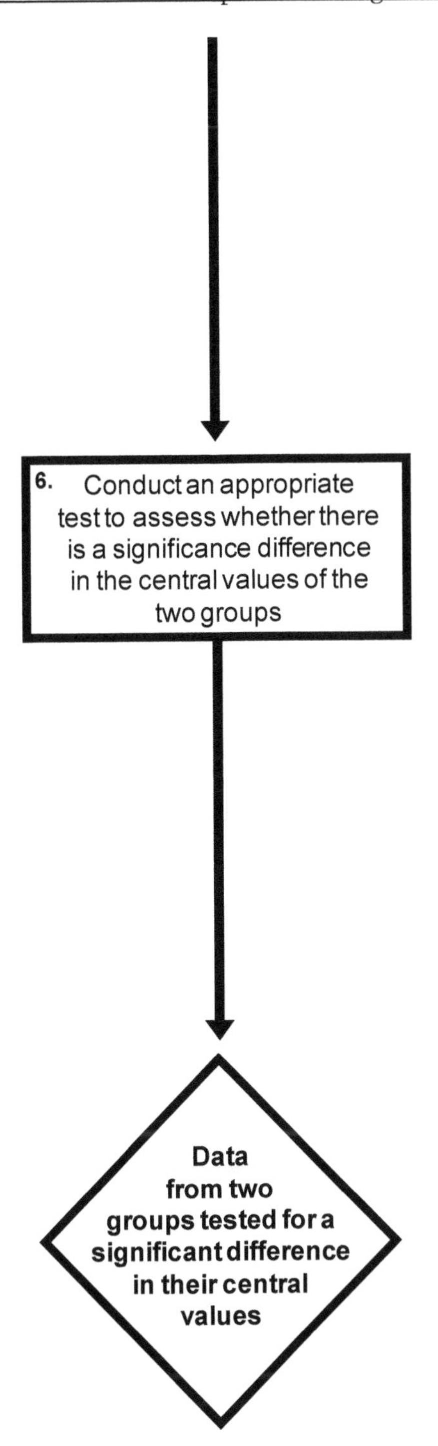
6. Conduct an appropriate test to assess whether there is a significance difference in the central values of the two groups
Data from two groups tested for a significant difference in their central values

Once you have determined which test is appropriate for your data set, you are ready to apply it. In this example, since neither the male nor the female mass data differed significantly from normal (see step 5), you can apply a t-test to assess whether or not there is a significant difference in mean body mass between the two groups. This is done on the R object containing the whole data set created in step 2 rather than on the R objects containing the data from each individual group created in step 5 (these were simply created and used for the normality tests). To do this, enter the following command into R:

```
t.test(great_tit_mass_data$mass~great_tit_
        mass_data$sex)
```

This is CODE BLOCK 72 in the document R_CODE_BASIC_STATS_WORKBOOK.DOC. This code applies a t-test to the mass data (held in the column called `mass`) from the data set called `great_tit_mass_data` based on the groupings provided in the column called `sex`.

At the end of the first part of this exercise, the box plot comparing the distributions of body mass values in male and female great tits should look like the image at the top of the next page.

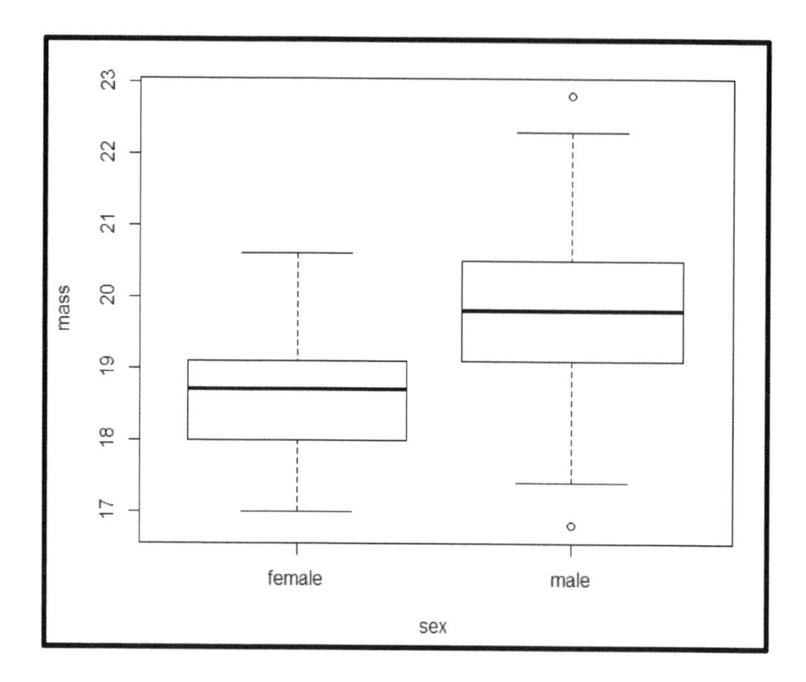

While the contents of your R CONSOLE window should look like this:

```
> t.test(great_tit_mass_data$mass~great_tit_mass_data$sex)

        Welch Two Sample t-test

data:  great_tit_mass_data$mass by great_tit_mass_data$sex
t = -5.5902, df = 84.72, p-value = 2.707e-07
alternative hypothesis: true difference in means is not equal to 0
95 percent confidence interval:
 -1.6539397 -0.7860603
sample estimates:
mean in group female   mean in group male
               18.61                19.83

> |
```

If you examine the results of this t-test, you will see that there is a significant difference in mean body mass between male and female great tits (p-value=2.707e-07 or 0.0000002707), with males being significantly heavier than females (female mean: 18.61g; male mean: 19.83g). To report the results of a t-test, you need to provide the value of the test statistic (t), the sample size (n) for each group, the degrees of freedom (d.f.), and the resulting p-value (p). In all cases, the values need to be rounded to an appropriate number of figures (see Appendix III for details). For the above t-test comparing mass in male and female great tits, you could report the results as outlined at the top of the next page.

When body masses of male and female great tits were compared, it was found that there was a significant difference in mean body mass between the two sexes (t-test: t =-5.6; n for males=50; n for females=50; d.f. =84.7; p<0.001). This means that, on average, male great tits are significantly heavier than females.

When comparing the central values between two groups using the above workflow, you can apply either a t-test to compare their means or a Mann-Whitney U test to compare their median values. Information on these two tests, along with the requirements of each test and the R command required to run them, are provided in the table below.

Test For Differences In Central Values	Test Requirements And How To Conduct It In R
T-Test	A t-test compares the means of a continuous variable for two independent groups. The distribution of the data for both groups must be normal. If the distribution of the data for either group is non-normal, you can try applying a transformation to normalise it before running your t-test (see Exercise 3.2). If you cannot normalise the data for one or both of your groups using a transformation, you should use a Mann-Whitney U test instead of a t-test (see below). To conduct a t-test in R, you can use the `t.test` command. To use this command, you need to define the column that contains the continuous variable and the column that contains the information about which group each data point belongs to. For example, to run this test on the contents of a column called `mass` based on groupings held in a column called `sex` in an R object called `great_tit_mass_data`, the full command would be: `t.test(great_tit_mass_data$mass~great_tit_mass_data$sex)` If you need to refine your t-test, there are a number of additional arguments that can be used with this command. Details of these additional arguments can be found at *www.rdocumentation.org/packages/stats/versions/3.6.1/topics/t.test.*
GLM-based T-Test	As well as using the `t.test` command, you can also conduct a t-test using the `glm` command. To conduct a t-test in R using this approach, you need to define the column that contains the continuous variable, the column that contains the group each data point belongs to and the name of the R object which contains these data. For example, to run a GLM-based t-test on the contents of a column called `mass` based on groupings held in a column called `sex` in an R object called `great_tit_mass_data`, the full command would be: `glm(mass~sex, data=great_tit_mass_data)` If you need to refine your GLM-based t-test, there are a number of additional arguments that can be used with this command. Details of these additional arguments can be found at *www.rdocumentation.org/packages/stats/versions/3.6.1/topics/glm.*

Test For Differences In Central Values	Test Requirements And How To Conduct It In R
Mann-Whitney U Test	A Mann-Whitney U test compares the medians of a continuous or ordinal variable for two independent groups. Since this is a non-parametric test, there is no requirement for the data to have normal distributions. To conduct a Mann-Whitney U test in R, you can use the `wilcox.test` command (**NOTE:** Despite the fact that you wish to run a Mann-Whitney U test, this is the name of the command you need to use to run this test in R). To use this command, you need to define the column that contains the continuous variable, the column that contains the group each data point belongs to and the R object which contains these data. For example, to run this test on the contents of a column called `mass` based on groupings held in a column called `sex` in an R object called `great_tit_mass_data`, the full command would be: `wilcox.test(mass~sex, data=great_tit_mass_data)` If you need to refine your Mann-Whitney U test, there are a number of additional arguments can be used with this command. Details of these additional arguments can be found at *www.rdocumentation.org/packages/stats/versions/3.6.1/topics/wilcox.test.*

For the next part of this exercise, you will customise the above approach for comparing the central values of data from two different groups in a number of different ways. Firstly, you will consider how you would have to modify the workflow in the above flow diagram if the distribution of the data from one or both of the groups is found to be non-normal. This will be done by comparing the masses of adult and juvenile great tits. To do this, you will need to modify the commands used in steps 4, 5 and 6.

For step 4, the modified versions of the R code should look like this (required modifications are highlighted in **bold**):

```
boxplot(mass~age, data=great_tit_mass_data)
```

For step 5, you would need to use the following modified versions of the subset commands to create separate data sets for adult and juvenile great tits:

```
adult_great_tit_data <- subset(great_tit_
        mass_data, age=="adult")
juvenile_great_tit_data <- subset(great_tit_
        mass_data, age=="juvenile")
```

You will also need to modify the code for running the normality tests on each subset of data in this step. The modified versions of this code should look like this:

```
shapiro.test(adult_great_tit_data$mass)
shapiro.test(juvenile_great_tit_data$mass)
```

At this stage, you will find that while the body mass data from juvenile great tits are normally distributed (p-value=0.1743), the body mass data from adults differ significantly from a normal distribution (p-value=0.00282). This means you will need to see if you can normalise the data using a transformation before you can proceed further with comparing the central values between the two groups. This can be done by entering the following command into R (see Exercise 3.2 for more information on conducting such transformations):

```
great_tit_mass_data$recip_mass=1/great_tit_mass_data$mass
```

This will add a new column (called `recip_mass`) to the R object called `great_tit_mass_data` and fill it with the reciprocally transformed body mass data. It is this transformed version of the body mass data that you will use for the rest of this example. Once you have done this, modify the R code for step 4 provided on page 138 to create a box plot based on the transformed version of the body mass data. The modified version of this code should look like this:

```
boxplot(recip_mass~age, data=great_tit_mass_data)
```

The box plot created by this command allows you to compare the distribution of the reciprocally transformed mass data for adult and juvenile great tits, and it should look like the image at the top of the next page.

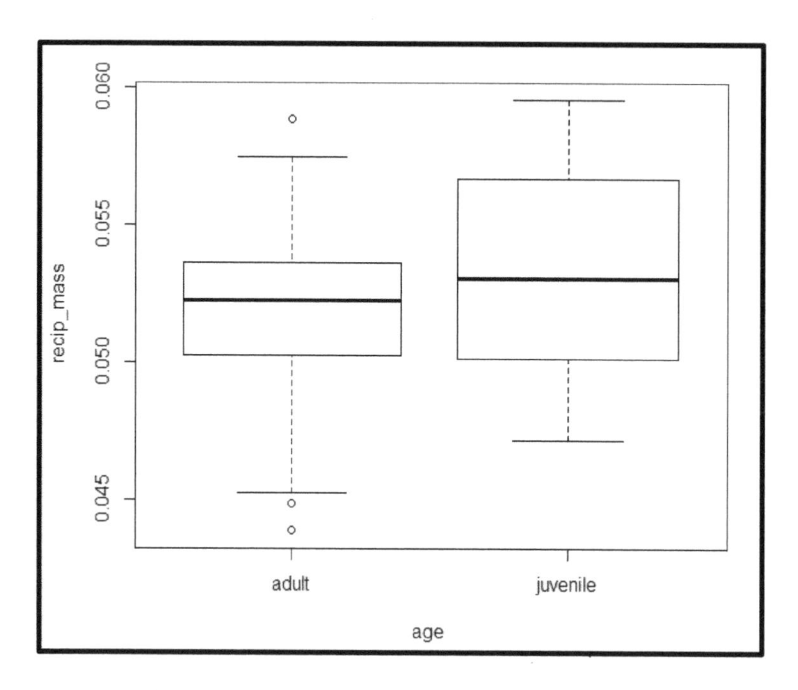

For step 5, you can use the same command you used on page 138 to create separate R objects for the adult and juvenile great tits again (**NOTE:** Even though you are using the same `subset` commands as before, this step has to be repeated as you have added a new column to the R object called `great_tit_mass_data` since you last ran it). However, when it comes to running the normality tests again, you will need to modify the code to reflect the fact that they now need to be run on the reciprocally transformed version of the body mass data (which is held in the column called `recip_mass` you just created). The modified versions of the code for running these normality tests should look like this:

```
shapiro.test(adult_great_tit_data$recip_mass)
shapiro.test(juvenile_great_tit_data$recip_mass)
```

At this stage, you will find that the reciprocally transformed data for both groups do not differ significantly from normal (p-value=0.1015 and p-value=0.1935 for adults and juveniles, respectively). This means that you can now move on to step 6 and use a t-test to compare the means of the data from these two groups. To do this, you will need to modify the code provided in step 6 to reflect that you are applying it to the reciprocally transformed body mass data from adults and juveniles, rather than the untransformed mass data from males and females (as was the case in the original workflow). The modified version of this code should look like the code at the top of the next page.

```
t.test(great_tit_mass_data$recip_mass~great_tit_
                    mass_data$age)
```

Once you have run this modified version of the `t.test` command, the contents of your R CONSOLE window should look like this:

```
> t.test(great_tit_mass_data$recip_mass~great_tit_mass_data$age)

        Welch Two Sample t-test

data:  great_tit_mass_data$recip_mass by great_tit_mass_data$age
t = -2.1163, df = 43.403, p-value = 0.04009
alternative hypothesis: true difference in means is not equal to 0
95 percent confidence interval:
 -3.187224e-03 -7.724873e-05
sample estimates:
   mean in group adult mean in group juvenile
            0.05178364             0.05341588

> |
```

If you examine the results of this t-test, you will see that there is a significant difference between the mean values of the reciprocally transformed body mass for adult and juvenile great tits, with adults, on average, having significantly heavier body masses than juveniles (mean reciprocally transformed body mass of adults: 0.05178364; mean reciprocally transformed body mass of juveniles: 0.05341588*). To report the results of this t-test, you would need to provide information about the transformation you used to normalise the data, the value of the test statistic (t), the sample size (n) for each group, the degrees of freedom (d.f.), and the resulting p-value (p). For the above t-test comparing mass in adult and juvenile great tits, you could report the results as follows:

When body masses of adult and juvenile great tits were compared, it was found that there was a significant difference in mean values of reciprocally transformed body mass between the two age groups (t-test: t=-2.1; n for adults=50; n for juveniles=50; d.f=43.4; p=0.040). This means that, on average, adult great tits are significantly heavier than juveniles.

*Since the formula for a reciprocal transformation is 1/(X) (see page 122 for details), a larger reciprocally transformed value is, rather confusingly, equivalent to a smaller untransformed value and *vice versa*. In this example, adult great tits have a significantly smaller mean reciprocally transformed body mass. As a result, this means that in absolute terms, their body mass is significantly greater than that of juveniles.

Once you have learned how to customise the workflow for conducting tests to compare the central values of two groups of data, you can test your ability to use it correctly by considering what you would need to do if, in step 5, you found that the data for one or both of your groups had a distribution that differed significantly from normal and that they could not be normalised using a transformation. This would mean that you would need to apply the non-parametric Mann-Whitney U test to compare the median values of the data from the two groups rather than comparing the means using a t-test. To explore how you would do this, you can compare the central values for the distributions of wing length between adult and juvenile great tits. To do this, you will need to modify the code used in steps 4, 5 and 6 when you compared the body mass of adult and juvenile great tits (see pages 138 – 141).

For step 4, the modified version of the `box plot` command should look like this:

```
boxplot(wing_length~age, data=great_tit_mass_data)
```

For step 5, you can use the `subset` command at the top of page 138 to once again create separate R objects for the adult and the juvenile great tits, but when it comes to running the normality tests, you will need to modify the code to reflect the fact that they are being run on data from a different column (`wing_length`). The modified versions of the code for running these normality tests should look like this:

```
shapiro.test(adult_great_tit_data$wing_length)
shapiro.test(juvenile_great_tit_data$wing_length)
```

At this stage, you will find that the wing length data for both adults and juveniles differ significantly from a normal distribution (p-value=4.261e-07 and p-value=0.008337 respectively). This means you will need to see if you can normalise the data using a transformation before you can proceed further with comparing the central values between the two groups. This can be done by entering the command at the top of the next page into R (see Exercise 3.2 for more information on conducting such transformations).

```
great_tit_mass_data$recip_wing_length=1/great_tit_
mass_data$wing_length
```

This will add a new column (called `recip_wing_length`) to the table of great tit data held in the R object called `great_tit_mass_data` and fill it with the reciprocally transformed wing length data. It is this newly transformed version of the wing length data that you will use for the next two steps of this example. Once you have done this transformation, repeat step 4 and create a box plot based on the transformed version of the wing length data. The modified versions of the code for creating this new box plot should look like this:

```
boxplot(recip_wing_length~age, data=great_tit_mass_data)
```

For step 5, you can use the original `subset` command from page 138 to create separate R objects for adult and juvenile great tits, but when it comes to running the normality tests again, you will need to modify the code to reflect the fact that they are being run on the reciprocally transformed wing length data. The modified versions of the code for running these normality tests should look like this:

```
shapiro.test(adult_great_tit_data$recip_wing_length)
shapiro.test(juvenile_great_tit_data$recip_wing_length)
```

At this stage, you will find that the data for both groups are still significantly different from normal (p-value=6.501e-07 and p-value=0.006322 for adults and juveniles, respectively). This means that you cannot apply a t-test to them in step 6, and instead you will need to use a use a Mann-Whitney U test (which, despite its name, is run in R using the `wilcox.test` command – see table on page 138 for details). This will be applied to the original, untransformed data for the variable being examined (since there is no requirement for these data to have a normal distribution). To do this, you would use the following code for step 6:

```
wilcox.test(wing_length~age, data=great_tit_mass_data)
```

Once you have run this command, the contents of your R CONSOLE window should look like the image at the top of the next page

```
> wilcox.test(wing_length~age, data=great_tit_mass_data)

	Wilcoxon rank sum test with continuity correction

data:  wing_length by age
W = 991.5, p-value = 0.9016
alternative hypothesis: true location shift is not equal to 0

> |
```

These results show that there is no significant difference in the median wing lengths of adult and juvenile great tits (p-value=0.9016). To report the findings of a Mann-Whitney U test, you would need to provide the value of the test statistic (W), the sample size (n), and the resulting p-value (p). For the above example, you could report it as follows:

When wing lengths of adult and juvenile great tits were compared, it was found that there was no difference between the median wing length between the two age groups (Mann-Whitney U test: W=991.5; n for adult great tits=50, n for juvenile great tits=50; p=0.902).

EXERCISE 4.2: HOW TO TEST FOR A DIFFERENCE IN THE CENTRAL VALUES (MEANS/MEDIANS) OF TWO PAIRED GROUPS:

In most cases, when you are testing for a significant difference in the means or medians (also known as central values) of data from two groups, all of your data points will be independent of each other (that is, the values of each data point have no influence on the values for any other data point). However, in some cases your data points will not be independent. For example, you may have two measurements from the same individual, one taken under experimental conditions and one taken under control conditions. Under such circumstances, the two measurements are not truly independent (because they come from the same individual) and this needs to be taken into account when comparing their central values. This is done by using what is known as a paired test. If the data from both groups of linked data have a normal distribution, you can use a paired t-test to investigate whether there is a significant difference in their mean values. If the data from one or both of your groups are non-normal, you can use a Wilcoxon Matched Pairs test to investigate whether there is a significant difference in their median values.

In order to be able to test for differences in the central values of two groups of linked data, you need to have your data in a spreadsheet or table with one row for each pair of linked data points. In this table, you also need to have one column containing the first measure of a continuous variable for each data point and another column containing the second linked measure for the same continuous variable. **NOTE:** This is a different data structure to the one recommended in Exercise 4.1 for conducting comparisons of the central values of independent or unlinked data.

For this exercise, you will test whether there are significant differences between individual male great tits for morphological variables measured at different times of year. You will start by comparing their body masses in January and July. To do this, work through the following flow diagram:

Data for analysis held in a comma separated value (.CSV) file

For this example, the data set you will use is stored in a file called `great_tit_seasonal_mass.csv` that is located in the WORKING DIRECTORY folder you created during the introduction to this chapter.

1. Set the WORKING DIRECTORY for your analysis project

Before you start any analysis in R, you first need to set the WORKING DIRECTORY. To do this, enter the text `setwd("` and then type the address of your WORKING DIRECTORY, using slashes (/) as the folder separators, before entering a second quotation mark followed by a closing bracket, like this `")`. For example, if your WORKING DIRECTORY has the address C:\STATS_FOR_BIOLOGISTS_ONE, your `setwd` command should look like this:

```
setwd("C:/STATS_FOR_BIOLOGISTS_ONE")
```

If you are using RGUI, enter your `setwd` command in the R CONSOLE window (remembering to use the address of your own WORKING DIRECTORY folder in it) and then press the ENTER key on your keyboard. If you are using RStudio, enter your `setwd` command into the SCRIPT EDITOR window. To run it, select it and then click on the RUN button at the top of this window. You will enter all the remaining commands for this exercise in a similar manner, depending on the user interface you are using.

To check that your WORKING DIRECTORY has been set properly, enter the command `getwd()` and carefully check that the address it returns is the same as the one for the STATS_FOR_BIOLOGISTS_ONE folder you created at the start of this chapter.

Before you move on to step 2, make sure that all the data you wish to use in your analysis project are located in this WORKING DIRECTORY folder. In this case, this is a file called `great_tit_seasonal_mass.csv`. **NOTE:** If the data you are going to import into R in step 2 are not located in the WORKING DIRECTORY you set in this step, the import code provided in the next step will not work.

2. Load your data into R using the `read.table` command

The `read.table` command provides the easiest way to load data held in a .CSV file into R so you can analyse it. To do this, you will use the following command:

```
great_tit_seasonal_mass <- read.table(file=
  "great_tit_seasonal_mass.csv", sep=",",
        as.is=FALSE, header=TRUE)
```

This code has to be entered exactly as it is written here or it will not work. If you wish to use the copy-and-paste approach for entering this command, copy the text directly below CODE BLOCK 73 in the document R_CODE_BASIC_STATS_WORKBOOK.DOC and paste it into R.

This command will create a new object in R called `great_tit_seasonal_mass` which will contain the data from the specified .CSV file. To load a different .CSV file into R, all you need to do is change the file name in the `file` argument to the name of the one you wish to import. In addition, you can use whatever name you wish for the object which will be created by this command. To do this, simply replace `great_tit_seasonal_mass` at the start of the first line of the above code with the name you wish to use for it. **NOTE:** If your .CSV data set uses a semicolon as the decimal separator, you would need to replace the `sep=","` argument with `sep=";"`.

3. Check the data have loaded into R correctly by checking the names of the columns and by viewing it

Whenever you import any data into R, you need to check that they have loaded correctly. First, you need to check that all the required columns are present in the R object you just created. To do this, enter the following command into R:

```
names(great_tit_seasonal_mass)
```

This is CODE BLOCK 74 in the document R_CODE_BASIC_STATS_WORKBOOK.DOC. This command will return the names used for each column in the R object created in step 2. For this example, the names should be `id`, `july_mass`, `jan_mass`, `july_wing_length` and `jan_wing_length`.

Next, you should view the contents of the whole table using the `View` command. This is done by entering the following code into R:

```
View(great_tit_seasonal_mass)
```

This is CODE BLOCK 75 in the document R_CODE_BASIC_STATS_WORKBOOK.DOC. This command will open a DATA VIEWER window where you can examine your data set and check that the correct data have been loaded into R.

4. Create a box plot to provide an initial subjective assessment of the distribution of your data

Once your data have been successfully imported into R, you are ready to start assessing whether or not the data from either of your linked groups have a distribution that differs significantly from normal. The first step in this process is to create a box plot from these data. Since the data from your two linked groups are held in different columns, you will need to use a slightly different version of the boxplot command used in step 4 of Exercise 4.1. To do this, enter the following code into R:

```
boxplot(great_tit_seasonal_mass$jan_mass,
    great_tit_seasonal_mass$july_mass,
        names=c("January"," July"))
```

This is CODE BLOCK 76 in the document R_CODE_BASIC_STATS_WORKBOOK.DOC. This command creates a graph with separate boxes for the data in the columns called jan_mass and july_mass in the R object called great_tit_seasonal_mass created in step 2 of this exercise. These boxes are labelled as January and July using the information provided in the names additional argument. This means this graph can be used to provide a subjective assessment of how the values for body mass from the two linked groups compare.

Before you can decide which test to apply to objectively compare the central values of your two linked groups, you need to run a separate normality test on the data from each group. If you find that the data from both linked groups have normal distributions, you can use a paired t-test to compare their mean values. If the distribution of the data from one or both linked groups is non-normal, the first thing you should do is try to normalise it using a mathematical transformation (see Exercise 3.2). Once you have applied your transformation you can then return to this workflow and repeat steps 4 and 5 for the transformed version of the variable. When you do this, if you find that the data for both linked groups are now normal, you can apply a paired t-test to them in step 6. If you cannot find a transformation that successfully normalises the data from both linked groups, you will need to use a Wilcoxon Matched Pairs test to compare the medians of the data from the two linked groups.

5. Assess whether or not each group of data in your data set has a normal distribution

For this example, as your data are already held in different columns of your data set (which is required for analysing data from linked groups), you do not need to divide the data into subsets before you can carry out an appropriate normality test (see Exercise 3.1). In this case, you will use the Shapiro-Wilk test. To apply this test to the male great tit body mass data from January, enter the following command into R:

```
shapiro.test(great_tit_seasonal_
        mass$jan_mass)
```

This is CODE BLOCK 77 in the document R_CODE_BASIC_STATS_WORKBOOK.DOC. When you do this, you will find that the distribution of the January body mass data for male great tits is not significantly different from normal (p-value=0.5607).

This test is then repeated for the second linked group of data (in this case, the data for the same great tits from July). To do this, enter the following command into R:

```
shapiro.test(great_tit_seasonal_
        mass$july_mass)
```

This is CODE BLOCK 78 in the document R_CODE_BASIC_STATS_WORKBOOK.DOC. When you do this, you will find that the distribution of the July body mass data for male great tits is also not significantly different from normal (p-value=0.3691).

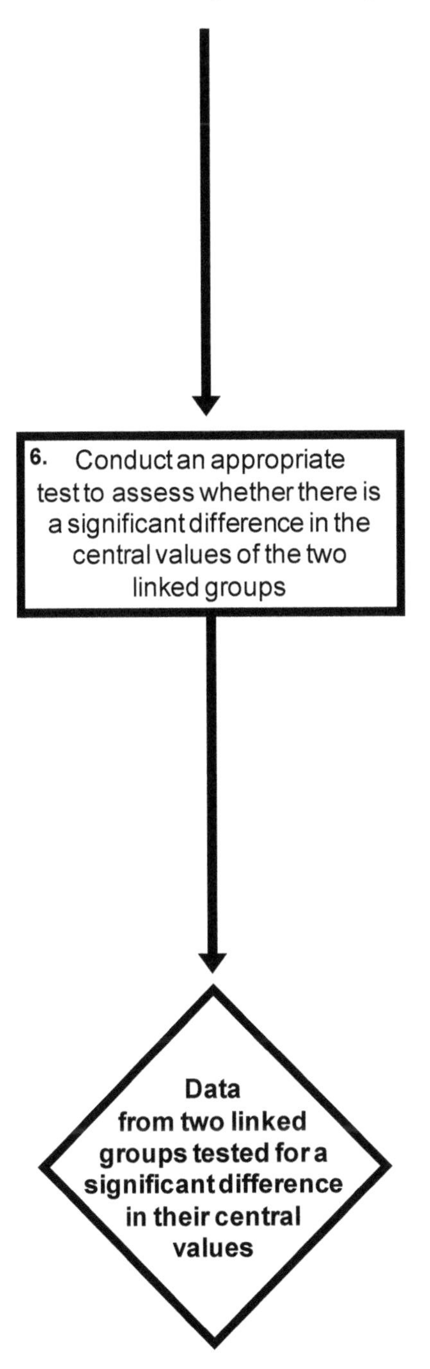
6. Conduct an appropriate test to assess whether there is a significant difference in the central values of the two linked groups
Data from two linked groups tested for a significant difference in their central values

Once you have determined which test is appropriate for your data set, you are ready to apply it. In this example, since neither the data on body masses from January nor from July differed significantly from normal (see step 5), you can apply a paired t-test to assess whether or not there is a significant difference in the mean mass between the two linked groups of data. To do this, enter the following command into R:

```
t.test(great_tit_seasonal_mass$jan_mass,
    great_tit_seasonal_mass$july_mass,
    paired=TRUE)
```

This is CODE BLOCK 79 in the document R_CODE_BASIC_STATS_WORKBOOK.DOC. This code applies a paired t-test (determined by the argument `paired=TRUE`) to the data held in the columns called `jan_mass` and `july_mass` in the R object called `great_tit_seasonal_mass`.

At the end of the first part of this exercise, the box plot comparing the distributions of body mass values in January and July from male great tits should look like the image at the top of the next page.

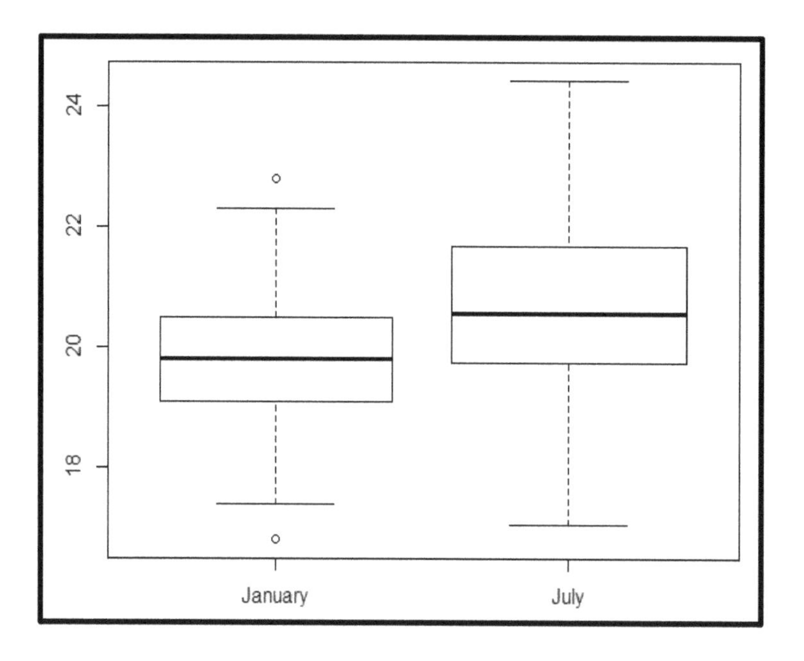

While the contents of your R CONSOLE window should look like this:

```
> t.test(great_tit_seasonal_mass$jan_mass, great_tit_seasonal_mass$july_mass, paired=TRUE)

        Paired t-test

data:  great_tit_seasonal_mass$jan_mass and great_tit_seasonal_mass$july_mass
t = -10.724, df = 49, p-value = 1.88e-14
alternative hypothesis: true difference in means is not equal to 0
95 percent confidence interval:
 -1.0970271 -0.7507649
sample estimates:
mean of the differences
              -0.923896

> |
```

If you examine the results of this paired t-test, you will see that there is a significant difference in mean body mass for the same male great tits from different months of the year, with their body mass being significantly greater in July than January (p-value=1.88e-14 or p-value=0.0000000000000188). The mean of the differences between the body masses in these two months is 0.923896g.

To report the results of a paired t-test, you need to provide the value of the test statistic (t), the sample size (n), the degrees of freedom (d.f.), and the resulting p-value (p). In all cases, these values need to be rounded to an appropriate number of figures (see Appendix III for details). For the above example, you could report it as follows:

When body masses of the same male great tits were compared in January and July, it was found that there was a significant difference between mean body mass of the same individuals from these two time periods (paired t-test: t=-10.7; n=50; d.f.=49; p<0.001). In the sample analysed, male great tits were, on average, heavier in July than in January.

When comparing the central values between two linked groups of data using the above workflow, you can apply either a paired t-test to compare their means or a Wilcoxon Matched Pairs test to compare their median values. Information on these two tests, along with the requirements of each test and the R command required to run them, are provided in the table below.

Test For Differences In Central Values	Test Requirements And How To Conduct It In R
Paired T-Test	A paired t-test compares the means of a continuous variable from two linked groups of data. The distribution of the data for both groups must be normal. If the distribution of the data for either group is non-normal, you can try applying a transformation to normalise it before running your paired t-test (see Exercise 3.2). If you cannot normalise the data for one or both of your linked groups using a transformation, you should use a Wilcoxon Matched Pairs test instead (see below). To conduct a paired t-test in R, you can use the `t.test` command with the argument `paired=TRUE`. To use this command, you need to define the columns that contain the first and second measures of the continuous variable for each data point in the linked groups. For example, to compare the mass of the same great tits measured in January (held in a column called `jan_mass`) and July (held in a column called `july_mass`) in an R object called `great_tit_seasonal_mass`, the full command would be: `t.test(great_tit_seasonal_mass$jan_mass, great_tit_seasonal_mass$july_mass, paired=TRUE)` If you need to refine your paired t-test, there are a number of additional arguments that can be used with this command. Details of these additional arguments can be found at *www.rdocumentation.org/packages/stats/versions/3.6.1/topics/t.test.*

Test For Differences In Central Values	Test Requirements And How To Conduct It In R
GLM-based Paired T-Test	As well as using the `t.test` command, you can also conduct a paired t-test using the `glm` command. To use this command, you need to define the column that contains the continuous variable, the column that contains the group each data point belongs to and the R object which contains these data. For example, to use a GLM-based paired t-test to compare the data in a column called `mass` based on the groups in a column called `month` for individuals identified by the contents of a column called `id` in an R object called `great_tit_paired_glm`*, the full command would be: `glm(mass~month+as.factor(id), data=great_tit_paird_glm)` If you need to refine your GLM-based paired t-test, there are a number of additional arguments that can be used with this command. Details of these additional arguments can be found at *www.rdocumentation.org/packages/stats/versions/3.6.1/topics/glm.*
Wilcoxon Matched Pairs Test	A Wilcoxon Matched Pairs test requires data for a continuous or an ordinal variable from two linked groups. Since this is a non-parametric test, there is no requirement for the data to have a normal distribution. To conduct a Wilcoxon Matched Pairs test in R, you can use the `wilcox.test` command with the additional argument `paired=TRUE`. To use this command, you need to define the columns that contain the first and second measures of the continuous variable for each data point in the linked groups, and the name of the R object which contains these data. For example, to use this test to compare the mass of the same great tits measured in January (held in a column called `jan_mass`) and July (held in a column called `july_mass`) in an R object called `great_tit_seasonal_mass`, the full command would be: `wilcox.test(great_tit_seasonal_mass$jan_mass, great_tit_seasonal_mass$july_mass, paired=TRUE)` If you need to refine your Wilcoxon Matched Pairs test, there are a number of additional arguments that can be used with this command. Details of these additional arguments can be found at *www.rdocumentation.org/packages/stats/versions/3.6.1/topics/wilcox.test.*

For the next part of this exercise, you will customise the above approach for comparing the central values of data from two linked groups. Specifically, you will consider how you would have to modify this workflow if the distribution of data for one or both of the linked groups is found to be non-normal. This will be done by comparing the central values of data on the wing length of male great tits recorded in January and July. To do this, you will need to modify the code used in steps 4, 5 and 6.

*The data structure required to run a paired t-test using the `glm` command is different from that required for other ways of doing a paired test. As a result, if you wish to try using this way of conducting a paired test, you will need to import a different data set (called `great_tit_paired_glm.csv`) into R.

For step 4, the modified versions of R code should look like this (required modifications are highlighted in **bold**):

```
boxplot(great_tit_seasonal_mas$jan_wing_length, great_
tit_seasonal_mass$july_wing_length, names=c("January",
                    "July"))
```

For step 5, to run the normality tests, you will need to modify the code to reflect the fact they are being run on data from different columns (the ones called `jan_wing_length` and `july_wing_length`) in the R object called `great_tit_seasonal_mass`. The modified versions of the code for running these normality tests should look like this:

```
shapiro.test(great_tit_seasonal_mass$jan_wing_length)
shapiro.test(great_tit_seasonal_mass$july_wing_length)
```

When you run these tests, you will find that the data from both linked groups have distributions that differ significantly from normal (p-values of 0.003507 and 7.111e-05 for the wing length data from January and July, respectively). At this stage, you would usually need to see if you can normalise the data using a mathematical transformation before you can proceed further with comparing the central values between the two linked groups. However, from Exercise 4.1 (which uses data from the same species of bird), we already know that the wing length data cannot easily be transformed to make it normal. Thus, for this example, you will move straight on to using a Wilcoxon Matched Pairs test, a non-parametric equivalent of a paired t-test. This test compares the median values of the data from the two linked groups rather than comparing the means (as is the case with a t-test). To do this, you will need to modify the code used in step 6 so that it looks like this:

```
wilcox.test(great_tit_seasonal_mass$jan_wing_length,
great_tit_seasonal_mass$july_wing_length, paired=TRUE)
```

Once you have completed this example, the contents of your R CONSOLE window should look like the image at the top of the next page.

```
> wilcox.test(great_tit_seasonal_mass$jan_wing_length, great_tit_seasonal_mass$july_wing_length, paired=TRUE)

        Wilcoxon signed rank test with continuity correction

data:  great_tit_seasonal_mass$jan_wing_length and great_tit_seasonal_mass$july_wing_length
V = 574, p-value = 0.5431
alternative hypothesis: true location shift is not equal to 0

> |
```

If you examine the results of this Wilcoxon Matched Pairs test, you will see there is no significant difference between the median values of wing length recorded for the same birds in January and July (p-value=0.5431), which is exactly what you would expect as any difference in measurements between the two time periods is most likely due to random measurement error rather than any directional change, as is the case with body mass. To report the results of a Wilcoxon Matched Pairs test, you need to provide the value of the test statistic (V), the sample size (n), and the resulting p-value (p). For the above example, you could report it as follows:

When wing lengths of the same male great tits were compared using measurements taken in January and July, it was found that there was no significant difference between the median wing lengths in the two time periods (Wilcoxon Matched Pairs test: V=574; n=50; p=0.543).

EXERCISE 4.3: HOW TO TEST FOR A DIFFERENCE IN THE VARIANCES OF TWO OR MORE GROUPS:

While assessing whether or not the distribution of your data is normal is one of the most important steps when you are working out which statistical tests you can apply to a specific data set, it is not the only consideration. For example, tests like ANOVA (see Exercise 4.4) and simple linear regression (see Exercise 5.2) not only require data for different groupings and variables to have normal distributions, they also require them to have similar levels of variance or spread of data (see Figure 3). This means that before you can conduct such statistical analyses, you not only need to test your data for normality, you also need to compare their levels of variance.

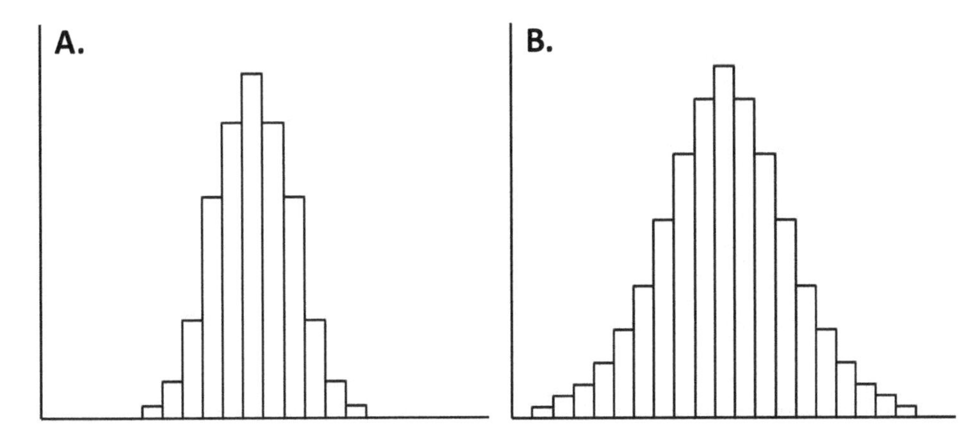

Figure 3: *The spread of data around the mean of a normal distribution is known as its variance. If two sets of data have different levels of variance around their respective means, like the data shown in Histograms A and B, it means that you cannot analyse them using tests like ANOVAs and simple linear regression.*

You can get an idea of whether or not two (or more) groups of data or the data for two (or more) variables have similar levels of variance by plotting a frequency distribution histogram (see Exercise 2.1) or a box plot (see Exercise 2.4). However, such assessments are subjective and, as a result, it is important that you also apply an objective test to assess whether or not your data violate any assumptions of equality of variances. There are two tests that biologists commonly use to do this. These are the F-test for Equality of Variances and Levene's test. Each of these tests has its own advantages and disadvantages when it comes to assessing whether two or more groups of data have similar levels of variance, but both of them require your data to have the same basic structure and that you work through the same basic steps.

In order to test whether your data have equal levels of variance, you need to have your data in a spreadsheet or table with one row for each data point in the data set, regardless of which group it belongs to. In this table, you also need to have a column containing the continuous variable you wish to compare between your groups and a second column which tells you which group each data point belongs to. For the F-test, as part of the workflow to carry it out, you will subset the data you wish to compare so the data from each group are held in different R objects, while for a Levene's test you do not need to subset your data in this way.

For this exercise, you will start by testing whether or not anglehead lizards from two locations in Malaysia have different levels of variance in body length. To do this, work through the following flow diagram:

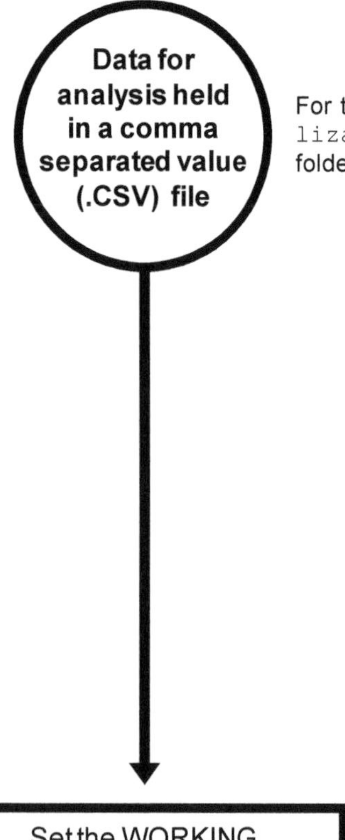
Data for analysis held in a comma separated value (.CSV) file
1. Set the WORKING DIRECTORY for your analysis project

For this example, the data set you will use is stored in a file called `lizad_data.csv` that is located in the WORKING DIRECTORY folder you created during the introduction to this chapter.

Before you start any analysis in R, you first need to set the WORKING DIRECTORY. To do this, enter the text `setwd("` and then type the address of your WORKING DIRECTORY, using slashes (/) as the folder separators, before entering a second quotation mark followed by a closing bracket, like this `")`. For example, if your WORKING DIRECTORY has the address C:\STATS_FOR_BIOLOGISTS_ONE, your `setwd` command should look like this:

```
setwd("C:/STATS_FOR_BIOLOGISTS_ONE")
```

If you are using RGUI, enter your `setwd` command in the R CONSOLE window (remembering to use the address of your own WORKING DIRECTORY folder in it) and then press the ENTER key on your keyboard. If you are using RStudio, enter your `setwd` command into the SCRIPT EDITOR window. To run it, select it and then click on the RUN button at the top of this window. You will enter all the remaining commands for this exercise in a similar manner, depending on the user interface you are using.

To check that your WORKING DIRECTORY has been set properly, enter the command `getwd()` and carefully check that the address it returns is the same as the one for the STATS_FOR_BIOLOGISTS_ONE folder you created at the start of this chapter.

Before you move on to step 2, make sure that all the data you wish to use in your analysis project are located in this WORKING DIRECTORY folder. In this case, this is a file called `lizard_data.csv`. **NOTE**: If the data you are going to import into R in step 2 are not located in the WORKING DIRECTORY you set in this step, the import code provided in the next step will not work.

2. Load your data into R using the `read.table` command

The `read.table` command provides the easiest way to load data held in a .CSV file into R so you can analyse it. To do this, you will use the following command:

```
lizard_data <- read.table(file="lizard_
    data.csv", sep=",", as.is=FALSE,
            header=TRUE)
```

This code has to be entered exactly as it is written here or it will not work. If you wish to use the copy-and-paste approach for entering this command, copy the text directly below CODE BLOCK 80 in the document R_CODE_BASIC_STATS_WORKBOOK.DOC and paste it into R.

This command will create a new object in R called `lizard_data` which will contain the data from the specified .CSV file. To load a different .CSV file into R, all you need to do is change the file name in the `file` argument to the name of the one you wish to import. In addition, you can use whatever name you wish for the object which will be created by this command. To do this, simply replace `lizard_data` at the start of the first line of the above code with the name you wish to use for it. **NOTE:** If your .CSV data set uses a semicolon as the decimal separator, you would need to replace the `sep=","` argument with `sep=";"`.

3. Check the data have loaded into R correctly by checking the names of the columns and by viewing it

Whenever you import any data into R, you need to check that they have loaded correctly. First, you need to check that all the required columns are present in the R object you just created. To do this, enter the following command into R:

```
names(lizard_data)
```

This is CODE BLOCK 81 in the document R_CODE_BASIC_STATS_WORKBOOK.DOC. This command will return the names used for each column in the R object created in step 2. For this example, the names should be `id`, `body_length`, `forelimb_length`, `location` and `sex`.

Next, you should view the contents of the whole table using the `View` command. This is done by entering the following code into R:

```
View(lizard_data)
```

This is CODE BLOCK 82 in the document R_CODE_BASIC_STATS_WORKBOOK.DOC. This command will open a DATA VIEWER window where you can examine your data set and check that the correct data have been loaded into R.

4. Create a box plot to provide an initial subjective assessment of the distribution of your data

Once your data have been successfully imported into R, you are ready to start assessing whether or not different groups in your data set differ significantly from each other. The first step in this process is to create a box plot of the data from the different groups. To do this, enter the following command into R:

```
boxplot(body_length~location, data=lizard_data)
```

This is CODE BLOCK 83 in the document R_CODE_BASIC_STATS_WORKBOOK.DOC. This command creates a box plot from the data in the column called `body_length` in the R object called `lizard_data` based on the categories in the column called `location`. This means it can be used to provide an initial subjective assessment of how the variances in the two groups of data compare.

Before you can decide which test to apply to objectively compare the variances of your groups of data, you first need to separate them into different data sets based on these groupings and then run a test of normality on each subset. This is because the F-test for Equality of Variances requires the data to have a normal distribution, while Levene's test does not. For the data being used in this example, you will divide the lizard data into two groups based on the contents of the `location` column using the `subset` command. To do this, enter the following code into R:

```
location_A <- subset(lizard_data,
        location== "A")
location_B <- subset(lizard_data,
        location=="B")
```

This is CODE BLOCK 84 in the document R_CODE_BASIC_STATS_WORKBOOK.DOC.

5. Assess whether or not each group of data in your data set has a normal distribution

Once you have created the subsets from your data based on the groupings you wish to compare, you can test each subset using an appropriate normality test (see Exercise 3.1). For the data being used in this example, you will use the Shapiro-Wilk test. To do this for the lizard body length data from location A, enter the following command into R:

```
shapiro.test(location_A$body_length)
```

This is CODE BLOCK 85 in the document R_CODE_BASIC_STATS_WORKBOOK.DOC. This will return a p-value of 0.4559, telling you that the lizard body length data from location A do not differ significantly from normal.

This test is then repeated for the second subset of data (in this case, the lizard data from location B). To do this, enter the following command into R:

```
shapiro.test(location_B$body_length)
```

This is CODE BLOCK 86 in the document R_CODE_BASIC_STATS_WORKBOOK.DOC. This will return a p-value of 0.9828, telling you that the lizard body length data from location B also do not differ significantly from normal.

If none of the groups in your data set have a distribution that differs significantly from normal (as is the case in this example), you can then use an F-test to compare the variances between them. If any of your subsets differ significantly from a normal distribution you will have to use a Levene's test instead.

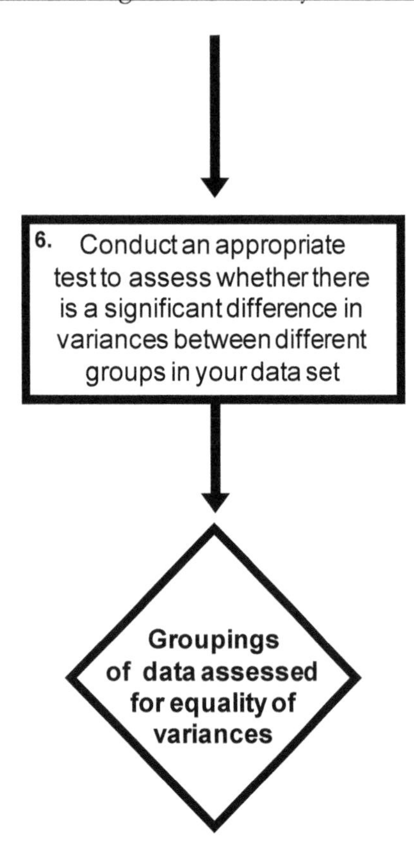

Once you have determined which test is appropriate for your data set, you are ready to apply it. Since neither the body length data from location A nor from location B differ significantly from normal, in this example you can apply an F-test test to assess whether or not these groups of data differ significantly from each other in terms of their variances. To do this, enter the following command into R:

```
var.test(location_A$body_length, location_B
         $body_length)
```

This is CODE BLOCK 87 in the document R_CODE_BASIC_STATS_WORKBOOK.DOC. This command applies an F-test to the data held in the columns called `body_length` in the R objects called `location_A` and `location_B` created in step 5.

At the end of the first part of this exercise, the box plot comparing the body lengths of anglehead lizards from locations A and B should look like this:

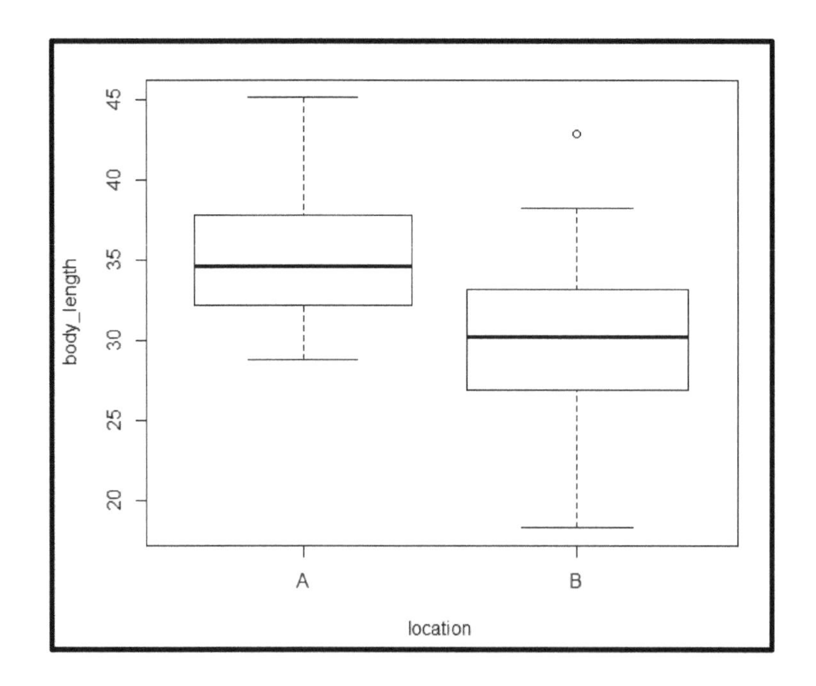

While the contents of your R CONSOLE window should look like this:

```
> var.test(location_A$body_length, location_B$body_length)

        F test to compare two variances

data:  location_A$body_length and location_B$body_length
F = 0.5146, num df = 19, denom df = 19, p-value = 0.1567
alternative hypothesis: true ratio of variances is not equal to 1
95 percent confidence interval:
 0.203685 1.300112
sample estimates:
ratio of variances
         0.5146001

> |
```

If you examine the results of this F-test, you will see there is no significant difference in the variances in lizard body lengths from location A and location B (p-value=0.1567). This means you could apply statistical analyses that require an equality of variance between groups to these data.

To report the results of a comparison of variances between groups of data, you need to provide details of the test used to compare the data, the test statistic (in this case, F), the degrees of freedom for each group (presented as a subscript to the test statistic), and the probability that any difference between the two samples is due to chance alone (p). In all cases, these values need to be rounded to an appropriate number of figures (see Appendix III for details). For the above example, you could report it as follows:

When the variances in body length in anglehead lizards sampled from two locations in Malaysia were compared, it was found that they did not differ significantly from each other (F-test: $F_{19,19}=0.52$; $p=0.157$).

When comparing variances between groups of data using the above workflow, you can apply either an F-test or a Levene's test in step 6. Information on these two variance tests, along with the requirements of each test and the R command required to run them, are provided in the table at the top of the next page.

Variance Test	Test Requirements And The Command Required To Run It In R
F-Test for Equality of Variances	An F-Test requires data for a continuous variable from two independent groups. The data for both groups must have a normal distribution. If the data from either group is non-normal, you can try applying a transformation to normalise it before running your F-test (see Exercise 3.2). If you cannot normalise the data for one or more of your groups using a transformation, you should use a Levene's test instead. To conduct an F-Test in R, you can use the `var.test` command. To use this command, you need to define the columns that contain the measures of the continuous variable for each of the two groups. These groups can either be in the same or in different objects within R. For example, to run this test to compare the variances in body lengths of lizards from location A (held in a column called `body_length` in an R object called `location_A`) and location B (held in a column called `body_length` in an R object called `location_B`), the full command would be: `var.test(location_A$body_length, location_B$body_length)` If you need to refine your F-test, there are a number of additional arguments that can be used with this command. Details of these additional arguments can be found at *www.rdocumentation.org/packages/stats/versions/3.6.1/topics/var.test.*
Levene's Test	A Levene's test requires data for a continuous variable from two or more independent groups. The distribution of the data from all these groups do not have to be normal. To conduct a Levene's test in R, you can use the `leveneTest` command. This command is contained in the `car` package, and you will need to install this package and library in R before you can use it (see page 166 for details). To use this command, you need to define the column that contains the measure of the continuous variable for all your groups, and the column that contains the information about which group each data point belongs to. For example, to run this test on the contents of a column called `body_length` in an R object called `lizard_data` based on the sampling location (from information held in a column called `location`), the full command would be: `leveneTest(lizard_data$body_length, lizard_data$location)` If you need to refine your Levene's test, there are a number of additional arguments that can be used with this command. Details of these additional arguments can be found at *www.rdocumentation.org/packages/car/versions/3.0-7/topics/leveneTest*

For the next part of this exercise, you will customise the above approach for comparing the variances of groups in a number of different ways. Firstly, you will run the same F-test again, but instead of comparing the variances of two groups of data, you will compare the variances of three groups. This is the modification you would need to make to the workflow outlined in the above flow diagram in order to assess whether you can apply an ANOVA to a specific data set. This will be done by comparing variances in body length of anglehead lizards from three different locations (the original two, and a new one, called location C). However, before you can do this, you will first need to import a new data set containing morphometric data from all three locations into R by modifying the `read.table` command from step 2, so that it looks like this (required modifications are highlighted in **bold**):

```
all_locations <- read.table(file="all_locations.csv",
        sep=",", as.is=FALSE, header=TRUE)
```

You can then work through the rest of the flow diagram, modifying the commands used at each step to reflect the fact that you are now working on a different data set with data from a different number of locations in it. For step 3, the modified commands should look like this:

```
names(all_locations)
View(all_locations)
```

For step 4, the modified versions of required R code should look like this:

```
boxplot(body_length~location, data=all_locations)
```

This will create a box plot comparing the variances in lizard body length between the three locations and it should look like this:

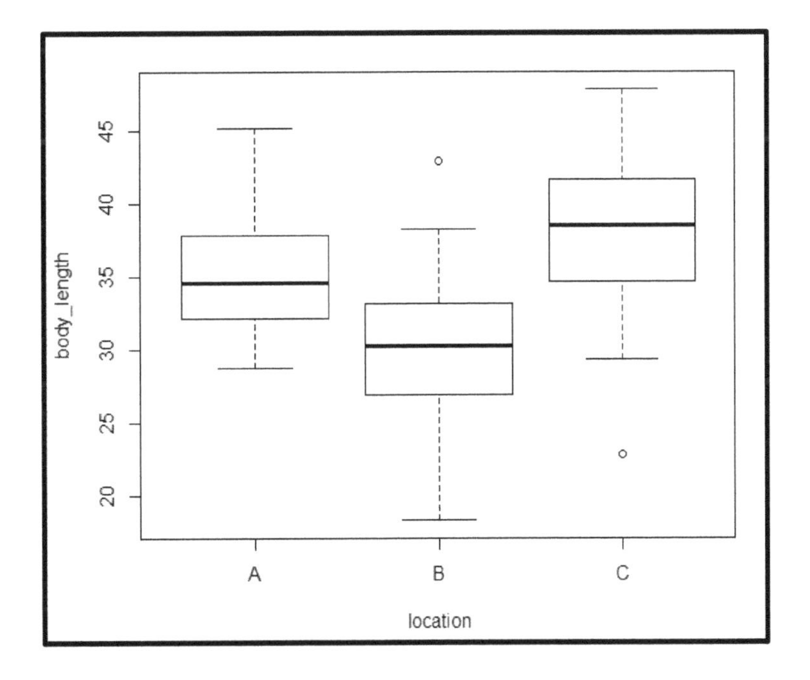

For step 5, the modified versions of the required commands should reflect the fact that you are now comparing data from three locations, not just two, and should look like this:

```
location_A <- subset(all_locations, location=="A")
location_B <- subset(all_locations, location=="B")
location_C <- subset(all_locations, location=="C")
```

```
shapiro.test(location_A$body_length)
shapiro.test(location_B$body_length)
shapiro.test(location_C$body_length)
```

And for step 6, the modified versions of the `var.test` commands should look like this:

```
var.test(location_A$body_length,location_B$body_length)
var.test(location_A$body_length,location_C$body_length)
var.test(location_B$body_length,location_C$body_length)
```

Once you have run all of the above modified commands, the contents of your R CONSOLE window should look like this:

```
> var.test(location_A$body_length, location_C$body_length)

        F test to compare two variances

data:  location_A$body_length and location_C$body_length
F = 0.43923, num df = 19, denom df = 19, p-value = 0.08082
alternative hypothesis: true ratio of variances is not equal to 1
95 percent confidence interval:
 0.1738527 1.1096940
sample estimates:
ratio of variances
         0.4392304

> var.test(location_B$body_length, location_C$body_length)

        F test to compare two variances

data:  location_B$body_length and location_C$body_length
F = 0.85354, num df = 19, denom df = 19, p-value = 0.7335
alternative hypothesis: true ratio of variances is not equal to 1
95 percent confidence interval:
 0.3378404 2.1564201
sample estimates:
ratio of variances
         0.8535373

> |
```

If you examine the results of these F-tests, you will see they found no difference in variances in body length between any pair of locations (A *vs* B: p-value=0.1567; A *vs* C: p-value=0.08082; B *vs* C: p-value=0.7335). This means you could apply statistical analyses that require an equality of variance between groups, such as an ANOVA, to these data.

When you have conducted multiple pair-wise comparisons between groups, you need to report the results of each individual comparison. For the above example, you could report them as follows:

When the variances in the body lengths of anglehead lizards from locations A, B and C were compared, it was found that it did not differ significantly between any pair of locations (location A vs location B: F-test: $F_{19,19}=0.51; p=0.157$; location A vs location C: F-test: $F_{19,19}=0.44; p=0.081$; location B vs location C: F-test: $F_{19,19}=0.85; p=0.734$).

Once you have learned how to customise the workflow for conducting tests for differences in variances between groups of data, you can test your ability to use it correctly by repeating it using a Levene's test. This will be applied to the same lizard body length data set from the same three locations that you used for the second set of F-tests. The main advantages of using a Levene's test over an F-test is that it does not require the data from all the groups to have a normal distribution, and that it allows multiple groups to be compared in a single test, rather than having to run multiple pair-wise comparison (as was the case for the above F-tests). To run a Levene's test on the lizard body length data from the three different locations, you would conduct steps 1 to 4 as outlined above for the `all_locations` data set. You can then skip step 5, as a Levene's test has no assumption of normality in the subsets being compared, while for step 6, rather than running multiple pair-wise comparisons of the body length data from different locations, you would compare the data from all three locations at once. Due to the differences in data structure required for Levene's test and the F-test, this is done by using the original data set and not the subsetted versions (as you would do for an F-test). To conduct a Levene's test on these data, enter the modified version of the command from step 6 shown at the top of the next page into R.

```
leveneTest(all_locations$body_length,
        all_locatons$location)
```

NOTE: If you get an error message saying that R cannot find the `leveneTest` command, the most likely reason for this is because you do not have the `car` library loaded into your R project. Before you can do this, you first need to install the `car` package by entering the following command into R:

```
install.packages("car")
```

Once this package has been installed, you can load the `car` library by entering the following command into R:

```
library(car)
```

You can then re-run the above version of the `leveneTest` command. After you have run it, the contents of your R CONSOLE window should look like this:

```
> install.packages("car")
Installing package into 'C:/Users/Computer/Documents/R/win-library/3.6'
(as 'lib' is unspecified)
trying URL 'https://cran.ma.imperial.ac.uk/bin/windows/contrib/3.6/car_3.0-6.zip'
Content type 'application/zip' length 1564112 bytes (1.5 MB)
downloaded 1.5 MB

package 'car' successfully unpacked and MD5 sums checked

The downloaded binary packages are in
        C:\Users\Computer\AppData\Local\Temp\Rtmpm42Z2x\downloaded_packages
> library(car)
Loading required package: carData

Attaching package: 'car'

The following object is masked from 'package:lawstat':

    levene.test

> leveneTest(all_locations$body_length,all_locations$location)
Levene's Test for Homogeneity of Variance (center = median)
      Df F value Pr(>F)
group  2  1.0145  0.369
      57
> |
```

If you examine the results of this Levene's test, you will see that (as with the above F-test comparisons on the same data) it found no significant difference in variance between any of the locations (Pr(>F)=0.369). To report the results of a Levene's test, you need to provide the value of the test statistic (F), the degrees of freedom (as a subscript to the test statistic), and the probability that the any differences between the groups being analysed are due to chance alone (p). For the above example, you could report it as follows:

When body length data from anglehead lizards sampled from three locations in Malaysia were compared, it was found that the variances did not differ significantly between any of the locations (Levene's Test: $F_{2,57}=1.10; p=0.369$).

EXERCISE 4.4: HOW TO TEST FOR DIFFERENCES BETWEEN THREE OR MORE GROUPS:

Most basic statistical tests, such as t-tests and Mann-Whitney U tests, are designed to compare data from two groups. While this makes them suitable for analysing data from many biological studies, a problem arises if you wish to compare data from three or more groups. To illustrate this issue, consider an experiment where you are investigating the impact of three different dietary regimes on body mass: a low carbohydrate diet, a low fat diet and a high protein diet. After feeding different groups of mice each of these diets for three months, you weigh them, giving you three sets of body mass data which will then need to be compared to see if there are any significant differences between them. Assuming your data are normally distributed, you could use a t-test to do three pair-wise body mass comparisons: The low carbohydrate group *vs* the low fat group; the low carbohydrate group *vs* the high protein group; and the low fat group *vs* the high protein group. The trouble with this is that you are doing multiple tests to investigate the same hypothesis (that following different dietary regimes can have a significant impact on body mass) with the same data set, and this can inflate the probability that you will find a significant difference by chance alone. As a result, it is better to use a single test to compare the data from all the different groups at once. The most common way that biologists do this is to run an ANOVA (which is short for Analysis of Variance). ANOVAs allow you to compare data from multiple groups in a single test, so avoiding any increased risk of finding a significant difference between groups by chance alone.

ANOVAs work by testing whether the differences between groups are greater than the variances within them. If they are, then this indicates that there are significant differences between the data from at least two of the groups being analysed. There are two types of ANOVA that biologists use on a regular basis. These are the parametric version, known simply as an ANOVA, and the Kruskal-Wallis One-Way Analysis of Variance test, which is its non-parametric equivalent. The parametric ANOVA assumes that the data within different samples are not only normally distributed (which can be tested with an appropriate normality test – see Exercise 3.1), but that they also all have similar levels of variance (which can be tested with an F-test or a Levene's test – see Exercise 4.3) and that they have similar sample sizes. If any of these assumptions are violated, it is not appropriate to use a parametric ANOVA, and instead the non-parametric Kruskal-Wallis test should be used. However, both tests assume that the data in different samples are independent of each other. While it is possible to conduct comparisons of data from three or more linked groups, this is beyond the scope of this introductory workbook.

In order to be able to use ANOVAs to test for significant differences in data from three or more groups, you need to have your data in a spreadsheet or table with one row for each data point in your data set for all of your groups. In this table, you also need to have a column containing the continuous variable you wish to compare between your groups and a second column containing the information that tells you which group each data point belongs to.

For this exercise, you will start by using an ANOVA to test if there are significant differences in the body lengths of anglehead lizards sampled from three locations in Malaysia. These are the same data that you used in Exercise 4.3 to test whether there was a significant difference in variances between different groups, and as you need to compare the variances of your data before you can conduct an ANOVA, you will use one of the R objects you created in Exercise 4.3 (called `all_locations`) as the basis for this exercise. This means that you need to have completed all of Exercise 4.3 before you can start this exercise. If you are doing an ANOVA on your own data, you will need to complete the workflow from the flow diagram in Exercise 4.3 to import your data, check it for normality and compare the variances in the different groups before you start on the workflow outlined below. To conduct your first ANOVA on the lizard data, work through the flow diagram that starts on the next page.

Data for analysis held in an existing object in R

For this example, the data set you will use is held in an R object called `all_locations` created as part of Exercise 4.3.

1. Conduct an ANOVA to assess whether there is a significance difference between your groups

Before you can conduct an ANOVA, you first need to set your WORKING DIRECTORY, import your data into R, check it, test it for normality and compare the variances of the data from the different groups. Details of how to do all of this can be found in Exercise 4.3, and it will provide you with all the information you need to determine whether you can use a parametric ANOVA, or whether you will need to use a non-parametric Kruskal-Wallis test.

For the lizard body length data being used in this example, you already know that the data from the three locations are normally distributed and do not have any significant differences in their variances (see Exercise 4.3). This means you can apply a parametric ANOVA test to them. There are three commonly used ways to run a parametric ANOVA in R: These are: Using the `aov` command, using the `Anova` command, and using the `glm` command (see table on pages 172-173). For this example, you will use the `aov` command. To do this for the lizard body length data, enter the following code into R:

```
lizard_anova <- aov(all_locations$body_
    length~all_locations$location)
```

This code has to be entered exactly as it is written here or it will not work. If you wish to use the copy-and-paste approach for entering this command, copy the text directly below CODE BLOCK 88 in the document R_CODE_BASIC_STATS_WORKBOOK.DOC and paste it into R. This code conducts an ANOVA on the data in the column called `body_length` based on the groupings provided in the column called `location` in the R object called `all_locations`. The results of this ANOVA are then saved in a new object in R called `lizard_anova`.

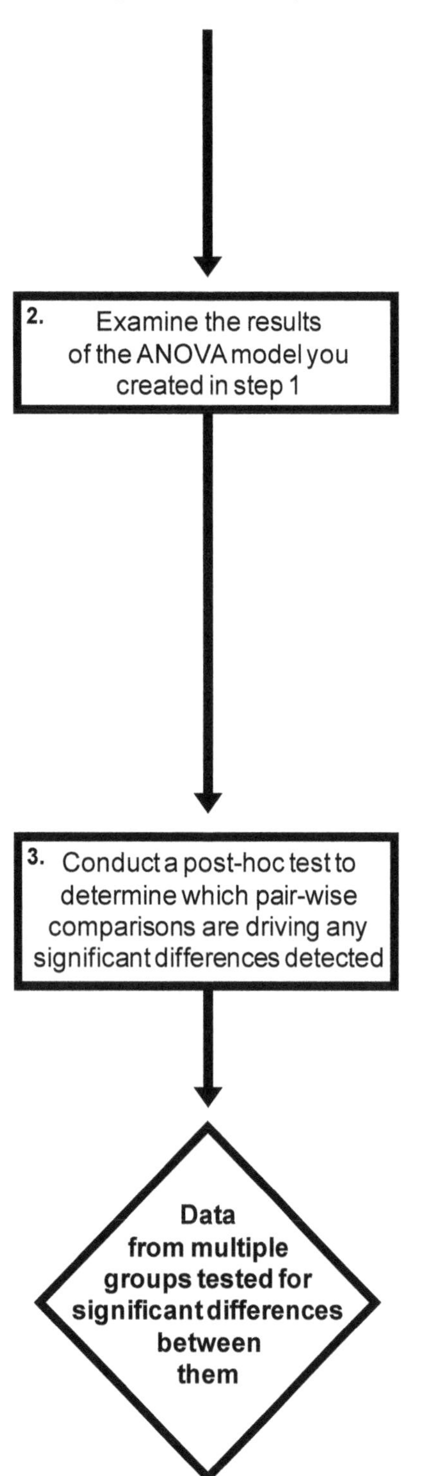
2. Examine the results of the ANOVA model you created in step 1
3. Conduct a post-hoc test to determine which pair-wise comparisons are driving any significant differences detected
Data from multiple groups tested for significant differences between them

Once you have created the new R object containing the results of your ANOVA, you can examine them. To do this, first enter the following command into R:

```
summary(lizard_anova)
```

This is CODE BLOCK 89 in the document R_CODE_BASIC_STATS_WORKBOOK.DOC. This command will create and display a summary table containing the results of your ANOVA.

Next, you can create a box plot to provide a visualisation of the pair-wise comparisons between the data analysed in your ANOVA. To do this, enter the following command into R:

```
boxplot(body_length~location,
    data=all_locations)
```

This is CODE BLOCK 90 in the document R_CODE_BASIC_STATS_WORKBOOK.DOC. This code will create a box plot that shows the distributions of the data in each grouping in your analysis.

When you create an ANOVA with the `aov` command, the table summarising its results (created by the `summary` command in step 2) tells you whether or not there is a significant difference between at least two of the groups within your data set. However, it does not provide any information about which pair-wise comparisons are significantly different from each other. This means that if your ANOVA finds a significant difference, you then need to conduct a further test to determine which pair-wise comparisons are driving this significant difference. This is done using the `TukeyHSD` command. To run this command on the ANOVA you conducted in step 1, enter the following code into R:

```
TukeyHSD(lizard_anova, conf.level=0.95)
```

This is CODE BLOCK 91 in the document R_CODE_BASIC_STATS_WORKBOOK.DOC.

At the end of the first part of this exercise, the box plot showing the distribution of the data in each of the groups analysed in your ANOVA should look like the image at the top of the next page.

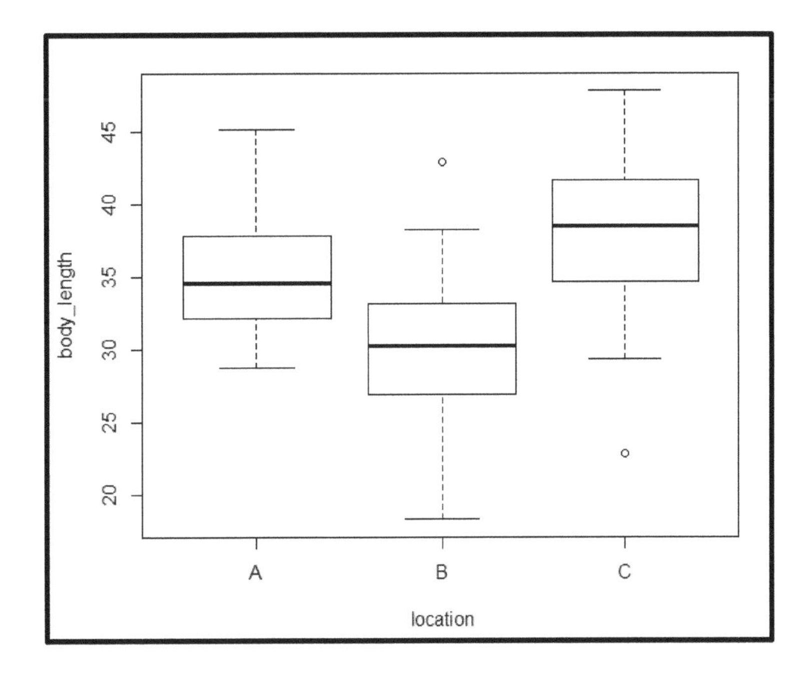

While the contents of your R CONSOLE window should look like this:

```
> lizard_anova <- aov(all_locations$body_length~all_locations$location)
> summary(lizard_anova)
                       Df Sum Sq Mean Sq F value   Pr(>F)
all_locations$location  2  575.2  287.60   10.48 0.000133 ***
Residuals              57 1564.4   27.45
---
Signif. codes:  0 '***' 0.001 '**' 0.01 '*' 0.05 '.' 0.1 ' ' 1
> boxplot(body_length~location,data=all_locations)
> TukeyHSD(lizard_anova, conf.level=0.95)
  Tukey multiple comparisons of means
    95% family-wise confidence level

Fit: aov(formula = all_locations$body_length ~ all_locations$location)

$`all_locations$location`
         diff       lwr        upr     p adj
B-A -4.739744 -8.726388 -0.7530992 0.0160071
C-A  2.757559 -1.229086  6.7442032 0.2276501
C-B  7.497302  3.510658 11.4839470 0.0000913

> |
```

If you examine the results of this ANOVA, you will see that there is a significant difference between the body lengths of anglehead lizards sampled from the three locations in Malaysia (Pr(>F)=0.000133). The ad-hoc Tukey test shows that this significant difference is driven by differences between the data from locations A and B (p adj=0.0160071) and locations B and C (p adj=0.0000913), while there was no significant difference between the data from locations A and C (p adj=0.2276501). To report the results of an ANOVA, you need to identify the test you used, the value of the test statistic (F), the degrees of freedom (presented as a subscript to the test statistic), the resulting p-value for the overall ANOVA as well as the results of any post-hoc tests, such as the Tukey test. In all cases, the values need to be rounded to an appropriate number of figures (see Appendix III for details). For the above example, you could report it as follows:

When body lengths of anglehead lizards from three locations in Malaysia were compared, it was found that there was a significant effect of location on body length (ANOVA: $F_{2,57}=10.5$; $p<0.001$). This location effect was driven by significant differences between lizards from locations A and B ($p=0.016$) and from locations B and C ($p<0.001$). The difference between lizards from locations A and C was not significant ($p=0.228$).

When comparing data from multiple groups using the above workflow, you can apply either a parametric ANOVA (which can be carried out as a traditional ANOVA or as a GLM-based ANOVA) or a non-parametric Kruskal-Wallis One Way ANOVA. Information on these tests, along with the requirements of each test and the R command required to run them, are provided in the table below

Test For Comparing Three Or More Groups	**Test Requirements And How To Conduct It In R**
ANOVA	An ANOVA requires data for a continuous variable from multiple independent groups. The data for all groups must have normal distributions and have similar levels of variance. If the data from any group is non-normal, you can try applying a transformation to normalise them before running your ANOVA (see Exercise 3.2). If you cannot normalise the data for one or more of your groups using a transformation, you should use a non-parametric Kruskal Wallis test instead. To conduct an ANOVA in R, you can use the `aov` command (for balanced data sets, like the one in the above example) or the `Anova` command (for unbalanced ones – this option will not be used in this workbook). To use the `aov` command, you need to define the column which contains the continuous variable as well as the one containing the information about which group each data point belongs to. For example, to run this test on the contents of a column called `body_length` based on groupings held in a column called `location` in an R object called `all_locations`, the full command would be: `aov(all_locations$body_length~all_locations$location)` If you need to refine your ANOVA, there are a number of additional arguments that can be used with the `aov` command. Details of these additional arguments can be found at *www.rdocumentation.org/packages/stats/versions/3.6.1/topics/aov*.

Test For Comparing Three Or More Groups	Test Requirements And How To Conduct It In R
GLM-based ANOVA	As well as using the `aov` and `Anova` commands, you can also conduct an ANOVA using the `glm` command. To conduct an ANOVA in R using this approach, you need to define the column that contains the continuous variable as well as the one that contains the group each data point belongs to and the name of the R object that contains these data. For example, to run a GLM-based ANOVA on the contents of a column called `body_length` based on groupings held in a column called `location` in an R object called `all_locations`, the full command would be: `glm(body_length~location, data=all_locations)` If you need to refine your GLM-based ANOVA, there are a number of additional arguments that can be used with this command. Details of these additional arguments can be found at *www.rdocumentation.org/packages/stats/versions/3.6.1/topics/glm.*
Kruskal-Wallis Test	A Kruskal-Wallis test requires data for a continuous or ordinal variable from multiple independent groups. There is no requirement for these data to have a normal distribution or have similar levels of variance. To conduct a Kruskal-Wallis test in R, you can use the `kruskal.test` command. To use this command, you need to define the column that contains the values of the variable being compared between the groups as well as the one containing the information about which group each data point belongs to and the name of the R object that contains these data. For example, to run this test on the contents of a column called `body_length` based on groupings held in a column called `location` in an R object called `all_locations`, the full command would be: `kruskal.test(body_length~location, data=all_locations)` If you need to refine your Kruskal-Wallis test, there are a number of additional arguments that can be used with this command. Details of these additional arguments can be found at *www.rdocumentation.org/packages/stats/versions/3.6.1/topics/kruskal.test.*

For the next part of this exercise, you will customise the above approach for comparing data from multiple groups in a number of different ways. Firstly, you will consider how you have to modify the workflow presented in the above flow diagram if you wish to run a GLM-based ANOVA rather than a traditional parametric ANOVA. This will be done using the same data set. As a result, the only thing you need to alter to run this second ANOVA is to change the R command in step 1. This is done by replacing the `aov` command with the `glm` command. The modified code for doing this should look like this (required modifications are highlighted in **bold**):

```
lizard_glm <- glm(body_length~as.factor(location),
                  data=all_locations)
```

You can then bring up the results of by adapting the command provided in step 2 so that it looks like this (**NOTE:** As well as using the `summary` command, for a GLM-based ANOVA, you also need to use a second command called `Anova` in order to bring up all the information required to assess the results of your analysis):

```
Anova(lizard_glm)
summary(lizard_glm)
```

At this stage of the exercise, the contents of your R CONSOLE window should look like this:

```
> lizard_glm <- glm(body_length~as.factor(location), data=all_locations)
> Anova(lizard_glm)
Analysis of Deviance Table (Type II tests)

Response: body_length
                    LR Chisq Df Pr(>Chisq)
as.factor(location)   20.957  2  2.813e-05 ***
---
Signif. codes:  0 '***' 0.001 '**' 0.01 '*' 0.05 '.' 0.1 ' ' 1
> summary(lizard_glm)

Call:
glm(formula = body_length ~ as.factor(location), data = all_locations)

Deviance Residuals:
     Min        1Q    Median        3Q       Max
-14.8326   -3.0202    0.2403    3.3948   12.7662

Coefficients:
                     Estimate Std. Error t value Pr(>|t|)
(Intercept)            34.873      1.171  29.769  < 2e-16 ***
as.factor(location)B   -4.740      1.657  -2.861  0.00589 **
as.factor(location)C    2.758      1.657   1.665  0.10150
---
Signif. codes:  0 '***' 0.001 '**' 0.01 '*' 0.05 '.' 0.1 ' ' 1

(Dispersion parameter for gaussian family taken to be 27.44561)

    Null deviance: 2139.6  on 59  degrees of freedom
Residual deviance: 1564.4  on 57  degrees of freedom
AIC: 373.93

Number of Fisher Scoring iterations: 2

> |
```

If you examine the results of this GLM-based ANOVA, you will see that there is significant difference in body length data from the three locations. As with the traditional parametric ANOVA, this is driven by a significant difference between the lizards from location B and the other two locations. To report the results of a GLM-based ANOVA, you need to provide the value of the test statistic (X^2), the degrees of freedom (d.f.), and the resulting p-value (p). For the above example, you could report it as follows:

When body lengths of anglehead lizards from three locations in Malaysia were compared, it was found that there was a significant effect of location on body length (GLM: $X^2=21.0$; d.f.=2; $p<0.001$). This effect is driven by differences in body length between individuals from location B and the other two locations.

If you wish to run a two-way ANOVA (that is, one which takes into account the combined, as well as individual, impacts of two factors on a continuous variable at the same time), this can be done by adding additional terms to the ANOVA command used in step 1. For example, for the anglehead lizard data, you might wish to check that any differences between locations are not being driven by difference in body size between males and females, especially if different numbers of each sex were sampled from each location. To do this, you would need to add a second variable (the sex of each individual sampled, which is contained in the column called `sex`) to your ANOVA model, as well as a term to allow it to look for an interaction between the two variables (`sex` and `location`). If you wish to do this using the `aov` command, the R code for doing it would look like this:

```
lizard_anova <- aov(all_locations$body_length~
all_locations$location + all_locations$sex +
all_locations$location*all_locations$sex)
```

If you wish to do this using the `glm` command, the R code for doing it would look like this:

```
lizard_glm <- glm(body_length~ as.facor(location)+
as.factor(sex)+as.factor(location)*as.factor(sex),
data=all_locations)
```

In both cases, the * symbol is used to indicate an interaction between two variables which have already been included individually within the model.

Once you have learned how to customise the workflow for conducting tests to compare data from three or more groups, you can test your ability to use it correctly by considering what you would need to do if you found that the data from different samples had significantly different levels of variance. This would be detected when applying the workflow from Exercise 4.3 prior to applying the ANOVA workflow presented in this exercise. To do this, you will test whether there are differences in values for a second morphological variable from the anglehead lizard data set. This variable is forelimb length. If you work through the flow diagram from Exercise 4.3 for this variable, you will find that while the data from each of the three locations are normally distributed, there is a significant difference in the variance between the data from locations A and B (p-value=0.02527). This means that when applying the ANOVA workflow, you would need to select the non-parametric Kruskal-Wallis One-Way ANOVA rather than the parametric ANOVA (as was used above). To apply a Kruskal-Wallis test to the forelimb length data from the three locations, you will need to modify the code used in steps 1 of the ANOVA workflow so that it looks like this:

```
lizard_forelimb_anova <- kruskal.test(forelimb_
    length~location, data=all_locations)
```

For step 2, the modified version of required R code should look like this:

```
lizard_forelimb_anova
boxplot(forelimb_length~location, data=all_locations)
```

The box plot created by the second part of this code should look like this:

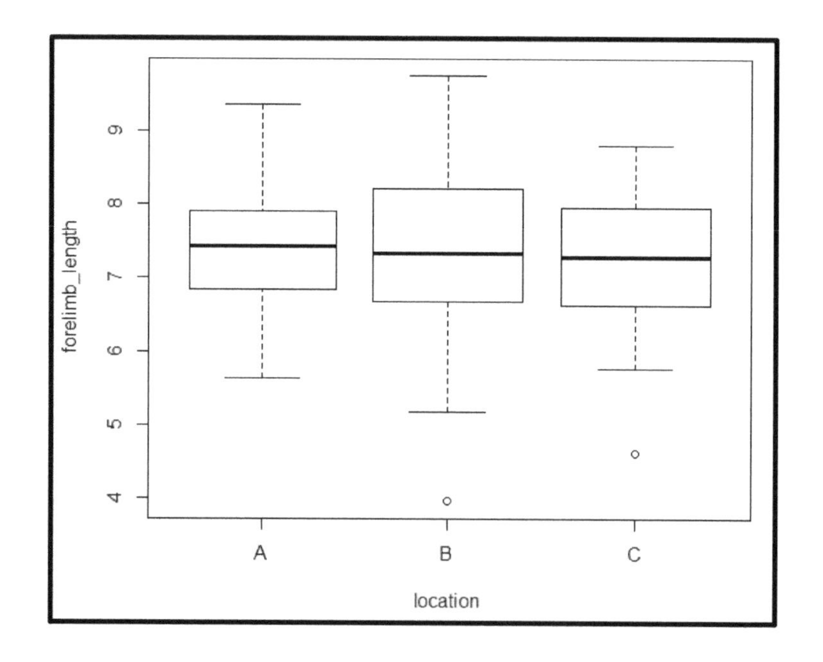

While the contents of your R CONSOLE window should look like this:

```
> lizard_forelimb_anova <- kruskal.test(forelimb_length~location,data=all_locations)
> lizard_forelimb_anova

        Kruskal-Wallis rank sum test

data:  forelimb_length by location
Kruskal-Wallis chi-squared = 0.38393, df = 2, p-value = 0.8253

> boxplot(forelimb_length~location,data=all_locations)
> |
```

If you examine the results of this Kruskal-Wallis test, you will see that there is no significant difference in forelimb length for anglehead lizards from the three sampling locations. As a result, there is no need to conduct a post-hoc test. If you did need to do this for a Kruskal-Wallis test, you would use multiple pair-wise Mann-Whitney U tests rather than a Tukey test. To report the results of a Kruskal-Wallis test, you need to provide the value of the test statistic (X^2), the degrees of freedom (d.f.), and the resulting p-value (p). For the above example, you could report it as follows:

When forelimb length of anglehead lizards from three locations were compared, it was found that there was no significant differences between them (Kruskal-Wallis test*: $X^2=0.38$; d.f.=2; p=0.825).*

EXERCISE 4.5 HOW TO TEST FOR DIFFERENCES IN FREQUENCIES OF OCCURRENCE MEASURED ON AN ORIDINAL OR CATEGORICAL SCALE:

So far, the tests that have been used in this chapter to compare data from different groups have required the variable being tested to have a more-or-less continuous distribution. However, there will be times when you wish to compare data from different groups that are recorded on ordinal or even categorical scales. For example, if you wish to investigate whether dark-coloured peppered moths really are more frequent in areas where tree trunks are also dark, you could count the number of light-coloured and dark-coloured moths in areas with trees with darker bark and in areas with trees with lighter bark. However, the resulting data set will consist of the counts of the number of moths in four categories: 1. Light-coloured moths in the areas with lighter bark; 2. Dark-coloured moths in the areas with lighter bark; 3. Light-coloured moths in the areas with darker bark; 4. Dark-coloured moths in the areas with darker bark. As a result of this data structure, it is not appropriate to compare the number of moths of each colour in the different environments using tests that compare central values (such as means in a t-test or medians in a Mann-Whitney u test) or variances (as is the case in F-tests and ANOVAs). Instead, you need to use a test which allows you to specifically compare the frequency of occurrence in different categories. The frequency of occurrence test most commonly used by biologists is the chi-squared test.

There are two basic ways that a chi-squared test can be used. These are: 1. To compare a recorded frequency of occurrence on an ordinal or categorical scale to an expected frequency of occurrence (which is known as a chi-squared goodness of fit test); 2. To compare the recorded frequency of occurrence between two or more groups to see if they come from the same underlying distribution (known as a chi-squared test of independence). For a goodness of fit test, the frequency of occurrence data need to be arranged in a table containing a single column with the number of records for each value on an ordinal scale or each group for a categorical variable. For tests of independence, the frequency of occurrence data need to be arranged in a table where the columns contain the categories which you wish to compare, while the rows contain the number of records for each of these categories. Thus, in both cases, the required data structure is very similar, and the only real difference is the number of columns present in the data set. However, in many cases, these data structures are very different from the tables or spreadsheets that biologists usually use to record their data. In such tables or spreadsheets, you will typically have a list of records with the information about which categories they belong to recorded in different columns. Thus, the first step in any chi-squared analysis is usually to summarise these data by the individual categories to create a table with the data structure required for analysis.

For this exercise, you will start by using a chi-squared goodness of fit test to investigate whether sea lice (an external parasitic arthropod) settle randomly on different parts of the bodies of Atlantic salmon or whether they preferentially settle on specific body parts. To collect data for this analysis, twenty Atlantic salmon were randomly sampled from fish farms and the number of immature sea lice (the infectious stage) were counted on the head, the fins, the body and the tail of each salmon. These numbers are provided in the table at the top of the next page. This table also includes two ways to measure the proportion of immature sea lice that would be expected to be recorded on the different body parts if there was no preferential selection for an individual body part during settlement.

Salmon Body Part	Recorded number of immature sea lice	Expected proportion based on an even number on each body part	Expected proportion based on relative size of each body part
Head	16	0.25	0.2
Fins	15	0.25	0.1
Body	14	0.25	0.6
Tail	15	0.25	0.1

For your first chi-squared test, you will use the proportion based on an even number settling on each body part as the expected probabilities in the goodness of fit test. To carry out this test, work through the following flow diagram:

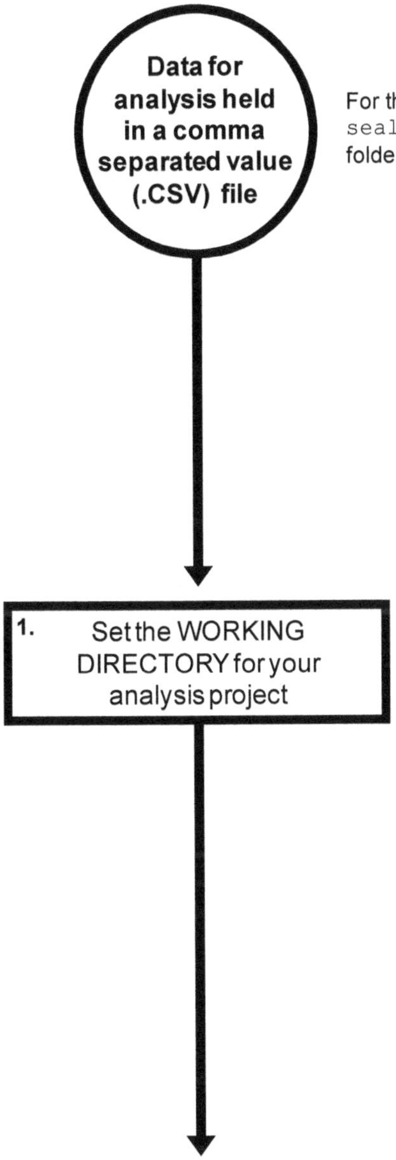

Data for analysis held in a comma separated value (.CSV) file

For this example, the data set you will use is stored in a file called `sealice_data.csv` that is located in the WORKING DIRECTORY folder you created during the introduction to this chapter.

1. Set the WORKING DIRECTORY for your analysis project

Before you start any analysis in R, you first need to set the WORKING DIRECTORY. To do this, enter the text `setwd("` and then type the address of your WORKING DIRECTORY, using slashes (/) as the folder separators, before entering a second quotation mark followed by a closing bracket, like this `")`. For example, if your WORKING DIRECTORY has the address C:\STATS_FOR_BIOLOGISTS_ONE, your `setwd` command should look like this:

```
setwd("C:/STATS_FOR_BIOLOGISTS_ONE")
```

If you are using RGUI, enter your `setwd` command in the R CONSOLE window (remembering to use the address of your own WORKING DIRECTORY folder in it) and then press the ENTER key on your keyboard. If you are using RStudio, enter your `setwd` command into the SCRIPT EDITOR window. To run it, select it and then click on the RUN button at the top of this window. You will enter all the remaining commands for this exercise in a similar manner, depending on the user interface you are using.

To check that your WORKING DIRECTORY has been set properly, enter the command `getwd()` and carefully check that the address it returns is the same as the one for the STATS_FOR_BIOLOGISTS_ONE folder you created at the start of this chapter.

Before you move on to step 2, make sure that all the data you wish to use in your analysis project are located in this WORKING DIRECTORY folder. In this case, this is a file called `sealice_data.csv`. **NOTE:** If the data you are going to import into R in step 2 are not located in the WORKING DIRECTORY you set in this step, the import code provided in the next step will not work.

2. Load your data into R using the `read.table` command

The `read.table` command provides the easiest way to load data held in a .CSV file into R so you can analyse it. To do this, you will use the following command:

```
sealice_data <- read.table(file="sealice_
    data.csv", sep=",", as.is=FALSE,
    header=TRUE)
```

This code has to be entered exactly as it is written here or it will not work. If you wish to use the copy-and-paste approach for entering this command, copy the text directly below CODE BLOCK 92 in the document R_CODE_BASIC_STATS_WORKBOOK.DOC and paste it into R.

This command will create a new object in R called `sealice_data` which will contain the data from the specified .CSV file. To load a different .CSV file into R, all you need to do is change the file name in the `file` argument to the name of the one you wish to import. In addition, you can use whatever name you wish for the object which will be created by this command. To do this, simply replace `sealice_data` at the start of the first line of the above code with the name you wish to use for it. **NOTE:** If your .CSV data set uses a semicolon as the decimal separator, you would need to replace the `sep=","` argument with `sep=";"`.

3. Check the data have loaded into R correctly by checking the names of the columns and by viewing it

Whenever you import any data into R, you need to check that they have loaded correctly. First, you need to check that all the required columns are present in the R object you just created. To do this, enter the following command into R:

```
names(sealice_data)
```

This is CODE BLOCK 93 in the document R_CODE_BASIC_STATS_WORKBOOK.DOC. This command will return the names used for each column in the R object created in step 2. For this example, the names should be `body_part`, `no_recorded`, `expected_even` and `expected_prop`.

Next, you should view the contents of the whole table using the `View` command. This is done by entering the following code into R:

```
View(sealice_data)
```

This is CODE BLOCK 94 in the document R_CODE_BASIC_STATS_WORKBOOK.DOC. This command will open a DATA VIEWER window where you can examine your data set and check that the correct data have been loaded into R.

4. Conduct an appropriate chi-squared test using the appropriate expected values

Once your data have been successfully imported into R, you are ready to run the appropriate version of the chi-squared test that you wish to use. In this example, you will run a chi-squared goodness of fit test. To run this type of chi-squared test, you need to provide details of the recorded distribution of your data and the expected distribution you wish to compare it to. In this case, you will compare the number of immature sea lice recorded on each body part of Atlantic salmon (this information is contained in the column called `no_recorded`) to the proportions expected if the same number of sea lice settled on each body part (this information is contained in the column called `expected_even`). To run this chi-squared goodness of fit test, enter the following command into R:

```
chisq.test(sealice_data$no_recorded,
    p=sealice_data$expected_even)
```

This is CODE BLOCK 95 in the document R_CODE_BASIC_STATS_WORKBOOK.DOC.

This `chisq.test` command compares the number of records in the column called `no_records` for different categories (represented by each line of data in the R object called `sealice_data`) to the value expected for that row. This value is calculated from the expected proportion provided in the column called `expected_even` from the same data set. Based on this comparison, this test returns the probability that any differences between the actual data set and the expected proportions in each category are due to chance alone.

5. Create a bar graph to visualise the comparison of the observed and the expected values for your data

Once you have run your chi-squared test, and determined if there is a significant difference between the observed and the expected values, it is useful to create a bar graph to visualise the comparison between them. To do this, you first need to create a new column called `no_expected_even` which contains the information on the expected number of records in each category (rather than the expected proportions). To do this, enter the following commands into R:

```
sealice_data$no_expected_even=
(sum(sealice_data$no_recorded))*sealice_data
$expected_even
View(sealice_data)
```

This is CODE BLOCK 96 in the document R_CODE_BASIC_STATS_WORKBOOK.DOC. Once you have created this new column, you can create a bar graph that compares the observed number of records in each category (contained in the column called `no_recorded`) with the number expected in each one (contained in the new column called `no_expected_even`). To create this bar graph, enter the following command into R (this is a modified version of the `barplot` command first introduced in Exercise 2.2):

```
barplot(rbind(sealice_data$no_recorded,
sealice_data$no_expected_even), beside=TRUE,
xlab="Body Part", ylab="Number of Immature
Sealice", col=c("white","black"),
names.arg=sealice_data$body_part,
ylim=c(0,20))
```

This is CODE BLOCK 97 in the document R_CODE_BASIC_STATS_WORKBOOK.DOC.

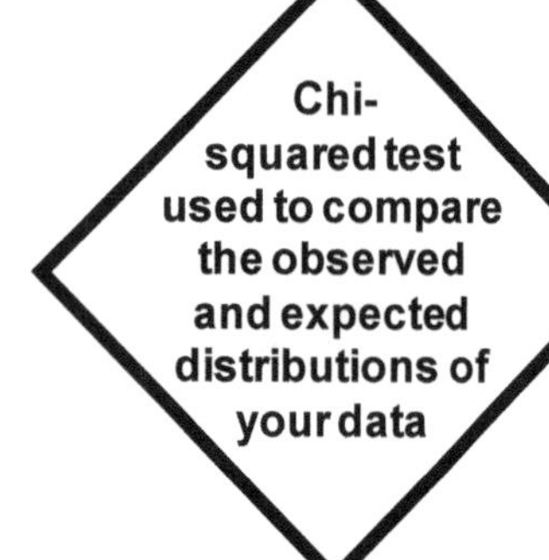
Chi-squared test used to compare the observed and expected distributions of your data

At the end of the first part of this exercise, the bar graph comparing your observed and your expected values should look like this:

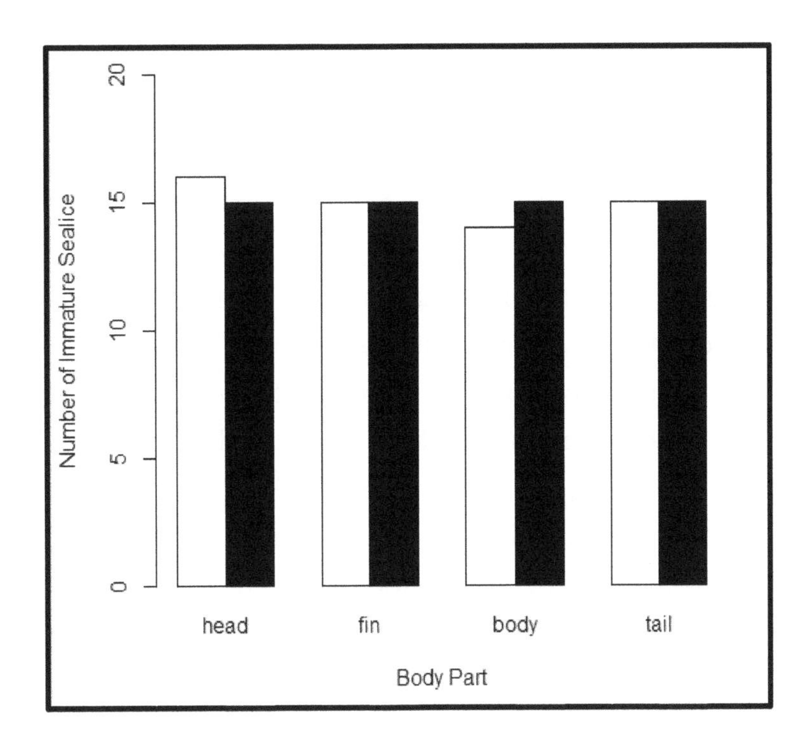

While the contents of your R CONSOLE window should look like this:

```
> sealice_data <- read.table(file="sealice_data.csv", sep=",", header=TRUE)
> names(sealice_data)
[1] "body_part"      "no_recorded"    "expected_even" "expected_prop"
> View(sealice_data)
> chisq.test(sealice_data$no_recorded, p=sealice_data$expected_even)

        Chi-squared test for given probabilities

data:  sealice_data$no_recorded
X-squared = 0.13333, df = 3, p-value = 0.9876

> sealice_data$no_expected_even= (sum(sealice_data$no_recorded))*sealice_data$expected_e
> barplot(rbind(sealice_data$no_recorded, sealice_data$no_expected_even), beside=TRUE, x
> 
```

If you examine the results of this chi-squared analysis, you will see that there is no significant difference between the observed and the expected values (p-value=0.9876). To report the results of this chi-squared test, you would need to provide the value of the test statistic (X^2), the degrees of freedom (d.f.), and the resulting p-value (p). In all cases, these values need to be rounded an appropriate number of figures (see Appendix III for details). For the above example, you could report it as follows:

When the number of immature sea lice recorded on four different parts of the bodies of farmed Atlantic salmon were compared to numbers expected based on an even distribution between them, there was no significant difference between the observed and the expected values (chi-squared goodness of fit test: $X^2=0.13$; d.f.=3; p=0.988).

However, while no significant difference was found in this first goodness of fit test, this does not necessarily mean that the settlement of sea lice on Atlantic salmon is random with respect to the different parts of the body. In particular, the assumption used to generate the expected values (that if settlement is random, you should expect an even number of immature sea lice on each body part) is not a particularly good assumption. This is because the individual parts do not necessarily make up the same proportion of a salmon's total surface area. For example, the body contains approximately 60% of a salmon's surface area, and so if settlement is random, you might expected 60% of sea lice to settle onto the it, and not the 25% used for the above chi-squared test. To assess the impact of this assumption on the findings of this analysis, it can be repeated using a different set of assumptions to calculate the expected values. Specifically, you can use the proportional area of each body part to generate them. This information is contained in the column called `expected_prop` in the R object called `sealice_data`. To re-run this chi-squared goodness of fit test using these different expected values, you would need to modify the `chisq.test` command in step 4 of the above flow diagram so that it looks like this (required modifications highlighted in **bold**):

```
chisq.test(sealice_data$no_recorded, p=sealice_
           data$expected_prop)
```

When you run this version of the chi-squared test, you will see that there is a significant difference between the observed and the expected values (p-value=4.472e-09 or p=0.00000000472). You can visualise this difference by plotting a bar graph comparing the observed and expected values used in this new test by modifying the commands provided in step 5 of the above flow diagram so that they look like the code at the top of the next page.

```
sealice_data$no_expected_prop=(sum(sealice_data$no_
recorded))*sealice_data$expected_prop
barplot(rbind(sealice_data$no_recorded, sealice_data$no_
expected_prop), beside=TRUE, xlab="Body Part", ylab="Number
of Immature Sealice", col=c("white","black"), names.arg=
sealice_data$body_part, ylim=c(0,40))
```

Once you have run this command, you should have a bar graph that looks like this:

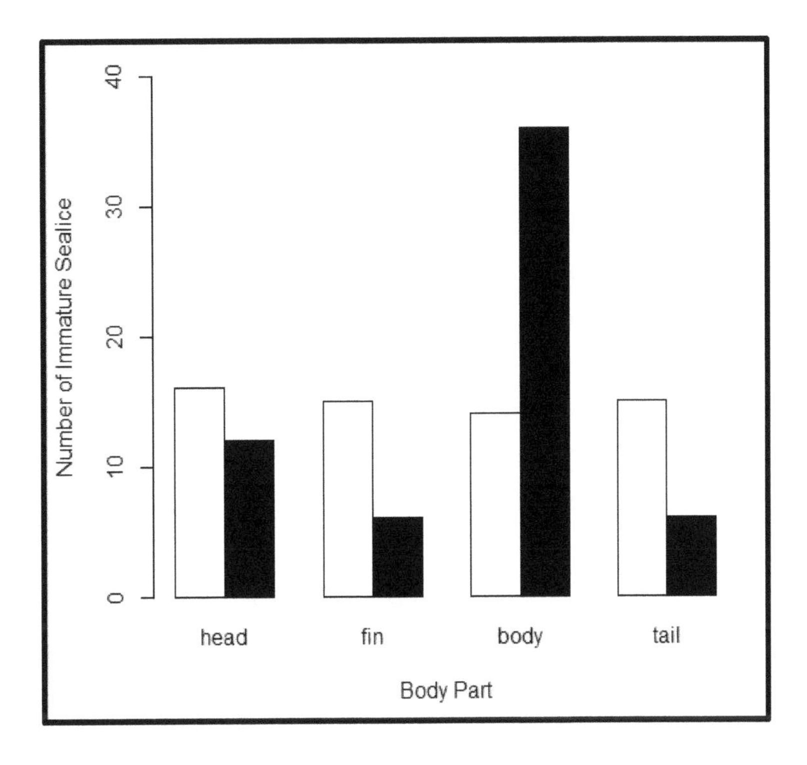

Running this second version of the test on the same data set demonstrates the importance of using an appropriate set of assumptions to calculate your expected values when conducting a chi-squared goodness of fit test. In fact, this is the most important decision you will have to make when applying this type of test to your own data.

As well as goodness of fit tests, you can also use chi-squared analysis to assess whether the data from multiple groups or samples have similar or different distributions. This is called a chi-squared test of independence. To gain experience in using this type of chi-squared test, you can modify the above workflow by changing how the data are processed prior to conducting the chi-squared test and how it is actually applied. To do this, you will examine whether the number of sea lice recorded on individual salmon has a similar distribution on fish sampled from fish farms located in sheltered and open environments. These data are held in a file called `fish_data.csv`. This data set has a different structure to the one you used for the above chi-squared goodness of fit test. In particular, rather than providing a table that summarises the information in your data set, the raw data are provided. In the case of the `fish_data.csv` data set, this means that each row contains the count of sea lice on an individual salmon (in a column called `no_lice`) along with information about the location of the fish farm which it came from (in a column called `environment`), and the sea lice prevalence category in a column called `lice_cat` (this assigns each fish to a specific category based on how many sea lice were recorded on it).

To run this analysis, you will first need to import these data into R by modifying the command provided in step 2 of the above flow diagram so that it looks like this:

```
fish_data <- read.table(file="fish_data.csv",
    sep=",", as.is=FALSE, header=TRUE)
```

Once these data have been imported into R, you need check they have been imported properly by modifying the commands in step 3 so that they look like this:

```
names(fish_data)
View(fish_data)
```

At this point, you are almost ready to run your chi-squared test of independence. However, before you can do this, you need to create something called a contingency table. This is a summary table that provides information about the number of records for each category you wish to analyse. For this example, your contingency table will provide a summary of the number of sampled fish with different numbers of sea lice from fish farms located in open and sheltered environments, and it should look like the table at the top of the next page.

Environment	0 Sealice	1 or 2 Sealice	3 or More Sealice
Open	20	6	1
Sheltered	4	16	39

This table can be generated in R from a list of data points using the `table` command. To do this for the `fish_data` data set, enter the following command into R:

```
contingency_table <- table(fish_data$environment,
                    fish_data$lice_cat)
```

Once you have generated this table, you can view it in R by entering its name without any additional commands (**NOTE:** If you use the `View` command to view a table generated by this `table` command, as you have done with other tables in this workbook, it will not show you the correct structure for your contingency table). To do this for the table you have just generated, enter the following code into R:

```
contingency_table
```

Your table will be displayed in your R CONSOLE window, and it should have the same layout as the one provided above. At this point, you are ready to run your chi-squared test of independence using the contingency table you just have generated from your data. This can be done by modifying the command provided in step 4 so that it looks like this:

```
chisq.test(contingency_table)
```

At this stage, the contents of your R CONSOLE window should look like this:

```
> contingency_table

            0 1 or 2 3 or more
  open     20      6         1
  sheltered 4     16        39
> chisq.test(contingency_table)

        Pearson's Chi-squared test

data:  contingency_table
X-squared = 45.738, df = 2, p-value = 1.17e-10

>
```

From this, you can see that there is significant difference between the number of sea lice recorded on Atlantic salmon from fish farms in open and sheltered environments (p-value=1.17e-10 or p=0.000000000117). To report the results of this chi-squared test, you would need to report the chi-squared value (X^2), the degrees of freedom (d.f.), and the resulting p-value (p). For the above example, you could report it as follows:

When the number of sea lice on Atlantic salmon sampled from fish farms in open and sheltered environments were compared to each other, there was a significant difference between the distribution of data across the three categories of sea lice prevalence (no sea lice, one or two sea lice, and three or more sea lice per individual salmon - chi-squared test: $X^2=45.7$; d.f.=2; $p<0.001$).

If you wish to visualise the comparisons between the prevalence of sea lice on Atlantic salmon from fish farms in open and sheltered environments, you would need to modify the R commands provided in step 5 of the above flow diagram. Before you can do this, you first need to separate your data into one R object that just contains the data from open environments and another that just contains the data for sheltered ones. This can be done for the above example by entering the following commands into R:

```
sheltered_environ <- subset(fish_data,
environment=="sheltered")
open_environ <- subset(fish_data, environment=="open")
```

Next, you need to create summary tables for each of these two subsets that can then be used to create a bar graph showing the required information. To do this for the above example, enter the following commands into R:

```
sheltered_graph <- aggregate(sheltered_environ$no_lice,
list(sheltered_environ$lice_cat),length)
colnames(sheltered_graph)=c("lice_cat","no_fish")
open_graph <- aggregate(open_environ$no_lice, list(open_
environ$lice_cat),length)
colnames(open_graph)=c("lice_cat","no_fish")
```

Once you have created these summary tables, you are ready to create the bar graph that will help you visualise the results of your chi-squared test of independence. To do this for the above example, enter the following command into R (this is a modified version of the `barplot` command provided in step 5 of the above flow diagram):

```
barplot(rbind(sheltered_graph$no_fish, open_graph$no_fish),
beside=TRUE, xlab="Number of Lice", ylab="Number of Fish",
col=c("white","black"), names.arg=sheltered_graph$lice_cat,
ylim=c(0,40))
```

Once you have run this command, you should have a bar graph that looks like this:

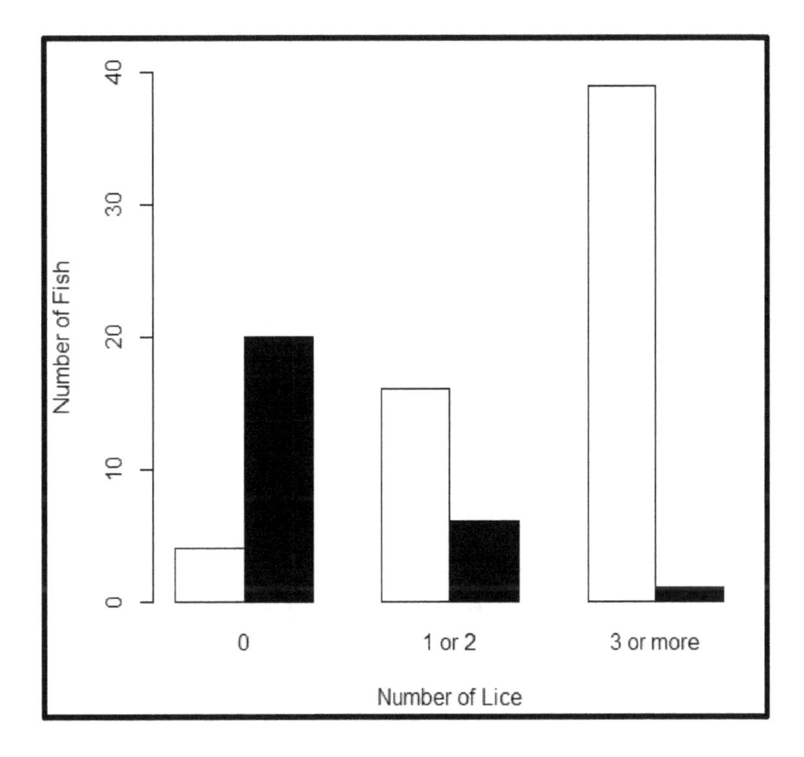

This graph shows that salmon from fish farms in sheltered environments (white bars) have a higher prevalence of sea lice than salmon from fish farms in more open environments (black bars).

--- Chapter Seven ---

Using Correlations And Regressions To Analyse Biological Data In R

In Chapter Six, you learned how to test for differences between the values of a specific variable in different groups. However, these tests can only be used to explore some types of biological relationships. For example, while they can be used to test whether males and females of the same species have different body masses, they cannot be used to tell you if there is a relationship between body mass and body length, and whether this relationship differs between males and females. It can be important to know about such relationships for two reasons. Firstly, it may be that relationships with other variables drive the differences between groups, and not the variable you are actually analysing. In the above example, it may be that in a particular species males are heavier than females simply because they are larger (as indicated by body length) rather than because they are heavier for their body size. As a result, it is important to know whether such relationships exist, and so whether they may influence the results of other types of analyses (meaning they need to be taken into account in some way). Secondly, it may be that to explore a specific research question, it is not differences between groups that you need to investigate, but rather it may be the relationships between variables themselves that are of interest. For example, you may wish to know if body size influences reproductive potential in fish. To explore such a research question, you would need to test whether there is a relationship between body size (as measured by morphological variables such as mass or length) and fecundity (as measured by the number of eggs a females produces).

There are two basic types of tests that biologists commonly use for exploring whether relationships exist between different variables. These are correlations and regressions. A correlation assesses whether there is a relationship between the values of two variables, but it does not provide any information about whether one variable is dependent on, or can be explained by, the other and, if so, which one is which. For example, a correlation analysis may find a relationship between body mass and body length, but it cannot tell you anything about which (if either) of these variables is driving this relationship. In contrast, a regression analysis specifically aims to determine the best relationship between a dependent or response variable and one (or more) independent or explanatory variables, and to use this relationship to estimate the most likely value of the response variable for any given value(s) of the explanatory variable(s). For example, a regression analysis might aim to determine whether body mass is dependent on body length and identify the exact relationship between the two. Thus, while a correlation does not provide any evidence of cause and effect, a regression specifically tests whether there is evidence that one variable is dependent on, or can be explained by, one (or more) other variables.

In this chapter, you will learn how to apply both correlations and regressions to biological data sets. As with other exercises in this workbook, you will start by importing your data into R and checking that the data are consistent with the required assumptions of a particular test before running the test itself, and ending with how to write up the results of your analysis.

If you have not already done so, before you start the exercises in this chapter, you first need to create a WORKING DIRECTORY folder on your computer and load the necessary data into it (**NOTE:** If you have already created this folder and downloaded data for a previous chapter in this workbook, you do not need to do this again). To do this on a computer with a Windows operating system, open Windows Explorer and navigate to the location where you would like to create the folder (such as your C:\ drive or your DOCUMENTS folder). Next, right click anywhere in this location and select NEW> FOLDER. Now call this folder STATS_FOR_BIOLOGISTS_ONE by typing this into the folder name section to replace what it is currently called (which will most likely be NEW FOLDER). To create a WORKING DIRECTORY folder on a computer running a Mac operating system, open Finder and navigate to the location where you would like to create the folder (such as your DOCUMENTS folder or your DESKTOP). Next, click on FILE> NEW FOLDER, and then type the name STATS_FOR_BIOLOGISTS_ONE before pressing the ENTER key on your keyboard.

Once you have created your WORKING DIRECTORY folder, you are ready to download the data sets you will use for the exercises in this workbook from *www.gisinecology.com/stats-for-biologists-1*. After you have downloaded the compressed folder containing the required data by following the instructions provided on that page, you need to extract all the data files

from it and copy them into the folder called STATS_FOR_BIOLOGISTS_ONE that you have just created.

Next, you need to check that the required data have been extracted to the correct folder. If you are using a computer with a Windows operating system, you can use Windows Explorer to open your newly created WORKING DIRECTORY folder and examine its contents. If all the files from the compressed folder are present in it (there should be a total of 21 of them), you can click on the folder icon at the left hand end of the ADDRESS BAR at the top of the WINDOWS EXPLORER window to reveal its full address. Write this address down as you will need it to set this folder as your WORKING DIRECTORY during the exercises provided in this workbook (see pages 12 and 13 for details of how to modify folder addresses so they will be recognised by R).

If you are using a computer with a Mac operating system, you can use Finder to open your newly created WORKING DIRECTORY folder and examine its contents. If all the required data files are present in it (there should be a total of 21 of them), select this folder in Finder and then press the CMD and I keys on your keyboard at the same time. This will open the GET INFO window where you will find its address (which is also called the pathway). Write this address down somewhere as you will need it to set this folder as your WORKING DIRECTORY during the exercises provided in this workbook (see pages 12 and 13 for details of how to modify folder addresses so they will be recognised by R).

After you have loaded the required data into your WORKING DIRECTORY folder, you can open RGUI or RStudio, depending on which option you wish to use (see Chapter 2 for more details). Once you have opened your preferred R user interface, you need to create a file called CHAPTER_SEVEN_EXERCISES where you will save the results of your analyses from your R CONSOLE window as you work through this chapter. To do this using RGUI, click on the FILE menu and select SAVE WORKSPACE. To do this in RStudio, click on SESSION and select SAVE WORKSPACE AS. In both cases, save it as a WORKSPACE file with the name CHAPTER_SEVEN_EXERCISES.RDATA in your WORKING DIRECTORY folder (this will be the one called STATS_FOR_ BIOLOGISTS_ONE that you have just created). If you are using RStudio, you will also want to save the contents of your SCRIPT EDITOR window (where you will enter and edit the R code you will use to carry out specific commands). To do this, click on the FILE menu and select SAVE AS. Save your file as an R SCRIPT file with the name CHAPTER_SEVEN_EXERCISES.R in your WORKING DIRECTORY folder. As you work through the exercises in this chapter, remember to regularly save the contents of your R CONSOLE window (which will contain the R objects you have created up to that point) to your WORKSPACE file and, if you are using RStudio, the contents of your SCRIPT EDITOR window to your R SCRIPT file.

Finally, you need to remove any data that are currently held in R's temporary memory. To do this, enter the following command into R (if you wish to copy and paste this command, the required code is directly below the text CODE BLOCK 1 in the document called R_ CODE_BASIC_STATS_WORKBOOK.DOC that is included in the compressed folder you just downloaded):

```
rm(list=ls())
```

If you are using RGUI, you can simply type or paste this code after the command prompt at the bottom of the R CONSOLE window (it looks like this: >) and then press the ENTER key on your keyboard to run it. If you are using RStudio, you can type or paste this command into the SCRIPT EDITOR window (the upper left hand window). To run this command, select it and then click on the RUN button at the top of this window. This will run it in the R CONSOLE window (the lower left hand one in the main RStudio user interface). You are now ready to start the exercises in this chapter.

EXERCISE 5.1: HOW TO TEST FOR A CORRELATION BETWEEN TWO VARIABLES:

Correlations are generally used in biology to investigate whether two variables are related. If there is a correlation between them, it means that as the value of one changes, so does the value of the other. As with almost all types of statistics, there are different ways of conducting a correlation depending the characteristics of the data you wish to analyse. The correlation tests most commonly used by biologists are the Pearson Product Moment Correlation Coefficient and the Spearman's Rank Correlation Coefficient. In both cases, these tests identify the best fit relationship between two variables and then measure how far away the actual data points are from this line of best fit. The closer the average distance between the points and this line of best fit, the stronger the correlation between the two variables. The strength of this relationship is measured using a correlation coefficient which will have a value between -1 and 1. A value of 0 indicates no correlation between the two, a value of 1 indicates a perfect positive correlation and a value of -1 indicates a perfect negative correlation.

The main differences between these two tests are that the Pearson Correlation Coefficient not only requires that the data for both variables are normally distributed, but also that they are measured on a continuous scale rather than a ordinal one and that the correlation between the two is linear. In contrast, the Spearman's Rank Correlation Coefficient does not require data to have a normal distribution, that the data are on a continuous scale, or that the correlation between the two variables is linear. In addition, as the correlation is based on the relative ranks of data points for two different variables, rather than actual values, it is less influenced by outliers within a data set. It does, however, assume that the relationship is monotonic (that is, it consistently goes up or down across the full range of data being analysed, and does not change from up to down or down to up at any point).

In order to investigate whether there is a correlation between two variables, you need to have your data in a spreadsheet or table with one row for each data point in your data set. In this table, you also need to have separate columns containing the values for each data point for the two variables you wish to compare. For this exercise, you will learn how to conduct a correlation analysis in R by examining how relative investment in reproduction varies with body mass in males from two mammalian taxa. These are the bats (Order Chiroptera) and the whales and dolphins (Order Cetacea). The research question for these analyses is whether, in general, males of larger species invest proportionately more or less in reproduction than males of smaller species. For both taxa, you will answer this question by investigating whether there is a correlation between the average male body mass for a species and its average percentage testes mass (this is the proportion of the total body mass that is accounted for by the testes). When comparing these variables, it is usual practice to log transform both variables prior to analysis, and this has already been done for you. If there is a positive correlation between the log-transformed values for these two variables, it means that males of larger species typically invest proportionately more in reproduction that males of smaller species (but it does not tell you whether being larger allows males to invest a higher proportion of their limited resources in reproduction, or whether males that need to invest a higher proportion of their resources in reproduction have to have a larger body size in order to be able to do this). If there is a negative correlation, it means that males of larger species typically invest proportionately less in reproduction. Finally, if there is no correlation, it means that there is no evidence of a relationship between the proportional investment in reproduction and body size.

The first possible relationship you will investigate will be whether there is a correlation between log-transformed male body mass and log-transformed percentage testes mass in microchiropteran bats (those species that have complex echolocation). To conduct this analysis, work through the following flow diagram:

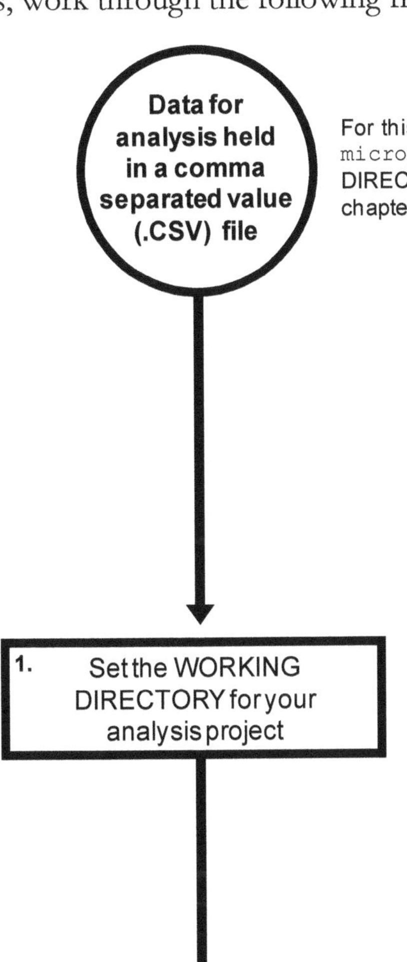
Data for analysis held in a comma separated value (.CSV) file
1. Set the WORKING DIRECTORY for your analysis project

For this example, the data set you will use is stored in a file called `micro_bat_data.csv` that is located in the WORKING DIRECTORY folder you created during the introduction to this chapter.

Before you start any analysis in R, you first need to set the WORKING DIRECTORY. To do this, enter the text `setwd("` and then type the address of your WORKING DIRECTORY, using slashes (/) as the folder separators, before entering a second quotation mark followed by a closing bracket, like this `")`. For example, if your WORKING DIRECTORY has the address C:\STATS_FOR_BIOLOGISTS_ONE, your `setwd` command should look like this:

```
setwd("C:/STATS_FOR_BIOLOGISTS_ONE")
```

If you are using RGUI, enter your `setwd` command in the R CONSOLE window (remembering to use the address of your own WORKING DIRECTORY folder in it) and then press the ENTER key on your keyboard. If you are using RStudio, enter your `setwd` command into the SCRIPT EDITOR window. To run it, select it and then click on the RUN button at the top of this window. You will enter all the remaining commands for this exercise in a similar manner, depending on the user interface you are using.

To check that your WORKING DIRECTORY has been set properly, enter the command `getwd()` and carefully check that the address it returns is the same as the one for the STATS_FOR_BIOLOGISTS_ONE folder you created at the start of this chapter.

Before you move on to step 2, make sure that all the data you wish to use in your analysis project are located in this WORKING DIRECTORY folder. In this case, this is a file called `micro_bat_data.csv`. **NOTE:** If the data you are going to import into R in step 2 are not located in the WORKING DIRECTORY you set in this step, the import code provided in the next step will not work.

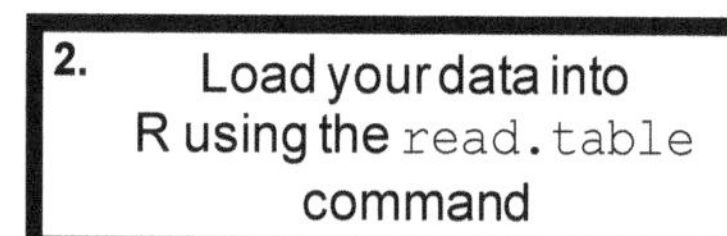

2.
Load your data into R using the read.table command

The `read.table` command provides the easiest way to import data held in a .CSV file into R so you can analyse it. To do this, you will use the following command:

```
micro_bat_data <- read.table(file="micro_
   bat_data.csv", sep=",", as.is=FALSE,
              header=TRUE)
```

This code has to be entered exactly as it is written here or it will not work. If you wish to use the copy-and-paste approach for entering this command, copy the text directly below CODE BLOCK 98 in the document R_CODE _BASIC_STATS_WORKBOOK.DOC and paste it into R.

This command will create a new object in R called `micro_bat_data` which will contain the data from the specified .CSV file. To load a different .CSV file into R, all you need to do is change the file name in the `file` argument to the name of the one you wish to import. In addition, you can use whatever name you wish for the object which will be created by this command. To do this, simply replace `micro_bat_data` at the start of the first line of the above code with the name you wish to use for it. **NOTE:** If your .CSV data set uses a semicolon as the decimal separator, you would need to replace the `sep=","` argument with `sep=";"`.

3. Check the data have loaded into R correctly by checking the names of the columns and by viewing it

Whenever you import any data into R, you need to check that they have loaded correctly. First, you need to check that all the required columns are present in the R object you just created. To do this, enter the following command into R:

```
names(micro_bat_data)
```

This is CODE BLOCK 99 in the document R_CODE_ BASIC_STATS_WORKBOOK.DOC. This command will return the names used for each column in the R object created in step 2. For this example, the names should be `id`, `log_body_mass`, `log_ptm` and `type`.

Next, you should view the contents of the whole table using the `View` command. This is done by entering the following code into R:

```
View(micro_bat_data)
```

This is CODE BLOCK 100 in the document R_CODE_ BASIC_STATS_WORKBOOK.DOC. This command will open a DATA VIEWER window where you can examine your data set and check that the correct data have been loaded into R.

4. Create a scatter plot to provide an initial subjective assessment of the relationship between your variables

Once your data have been successfully imported into R, you are ready to start assessing whether the relationship between the two variables you wish to use in your correlation is likely to be linear or non-linear (that is, whether the relationship between them can best be described by a straight line or whether it is better described by a curved one). The linearity of the relationship between the variables you are interested in can be subjectively explored by creating a scatter plot from them. To do this for the data set being analysed in this example, enter the following command into R:

```
plot(micro_bat_data$log_body_mass, micro_
        bat_data$log_ptm)
```

This is CODE BLOCK 101 in the document R_CODE_BASIC_STATS_WORKBOOK.DOC. This `plot` command will create a scatter plot from the data in the columns called `log_body_mass` and `log_ptm` in the R object called `micro_bat_data` created in step 2. This graph will allow you to subjectively assess whether the relationship between the two variables is linear enough to apply the parametric Pearson Product Moment Coefficient to it. If it is clearly non-linear, you will have to use the non-parametric Spearman's Rank Correlation Coefficient instead. The linearity of the relationship between a pair of variables can be subjectively assessed by holding a pen or pencil so that it passes through the middle of the points on your graph. If the relationship is broadly linear, there should be roughly the same number of points above and below the pen/pencil along its full length. If there are any parts of the graph where there are many more points on one side of it than the other, this suggests the relationship is non-linear. When you do this for the data being used in this example, you will find that the relationship is, indeed, linear enough to allow a Pearson Product Moment Correlation Coefficient to be conducted upon them.

As well as requiring that there is a linear relationship between the two variables being examined, calculating the parametric Pearson Product Moment Correlation Coefficient also requires that the data have a bivariate normal distribution. This is difficult to test for, but you can get an idea of whether this is likely to be the case by examining whether the distribution of the values for each variable is normally distributed. This can be done using a test for normality, such as the Shapiro-Wilk test (see Exercise 3.1 for more details). To apply this test to the log-transformed body mass data, enter the following command into R:

```
shapiro.test(micro_bat_data$log_body_mass)
```

5. Assess whether or not the data for each variable differ significantly from normal

This is CODE BLOCK 102 in the document R_CODE_BASIC_STATS_WORKBOOK.DOC. When you do this, you will find that the distribution of the log-transformed body mass data (contained in the column called `log_body_mass`) for microchiropteran bats is not significantly different from normal (p-value=0.06545).

This test should then be repeated for the second variable that you wish to use in your correlation (in this case, log-transformed percentage testes mass). To do this, enter the following command into R:

```
shapiro.test(micro_bat_data$log_ptm)
```

This is CODE BLOCK 103 in the document R_CODE_BASIC_STATS_ WORKBOOK.DOC. When you do this, you will find that the distribution of the log-transformed percentage testes mass data (contained in the column called `log_ptm`) for microchiropteran bats is also not significantly different from normal (p-value=0.1773).

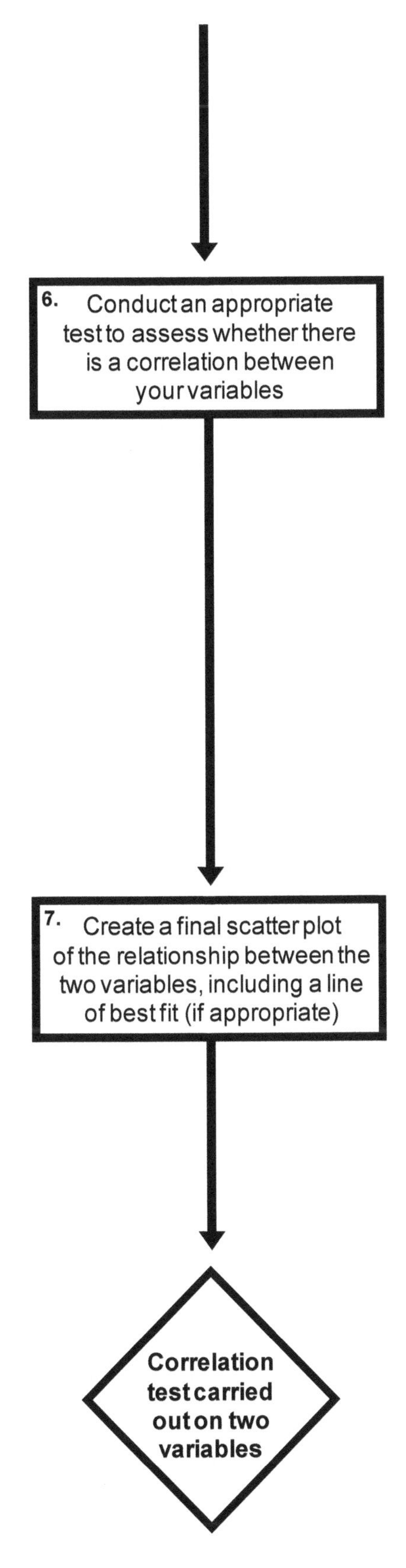
6. Conduct an appropriate test to assess whether there is a correlation between your variables
7. Create a final scatter plot of the relationship between the two variables, including a line of best fit (if appropriate)
Correlation test carried out on two variables

Once you have determined which test is appropriate for your data set, you are ready to apply it. In this example, since the relationship between log-transformed body mass and log-transformed percentage tests mass in microchiropteran bats looks like it is linear and the distributions of the two variables do not differ significantly from normal, you can use a Pearson Product Moment Correlation Coefficient to test whether there is a correlation between them. To do this, enter the following command into R:

```
cor.test(micro_bat_data$log_body_mass,
micro_bat_data$log_ptm, method="pearson")
```

This is CODE BLOCK 104 in the document R_CODE_BASIC_STATS_WORKBOOK.DOC. This code will calculate a Pearson Product Moment Correlation Coefficient from the data contained in the `log_body_mass` and `log_ptm` columns in the R object called `micro_bat_data`. To calculate a Spearman's Rank Correlation Coefficient, rather a Pearson Product Moment Correlation Coefficient, you can set the `method` argument to `method="spearman"` rather than `method="pearson"`.

You can now create a final scatter plot to visualise the correlation between your variables. To do this for the data being used in this example, enter the following command into R:

```
plot(micro_bat_data$log_body_mass, micro_
bat_data$log_ptm, xlab="Log Body Mass",
ylab="Log Percentage Testes Mass")
```

This is CODE BLOCK 105 in the document R_CODE_BASIC_STATS_WORKBOOK.DOC. This command creates a scatter plot from the data in the columns called `log_body_mass` and the `log_ptm` in the R object called `micro_bat_data`. It also adds appropriate labels to the X and Y axes.

If you have found that there is a significant linear correlation between the two variables, you can also add a line of best fit to your scatter plot. To do this for the data being used in this example, enter the following command into R:

```
abline(lm(micro_bat_data$log_ptm~micro_
bat_data$log_body_mass))
```

This is CODE BLOCK 106 in the document R_CODE_BASIC_STATS_WORKBOOK.DOC

The final scatter plot showing the correlation between log-transformed body mass and log-transformed percentage testes mass for microchiropteran bats should look like this:

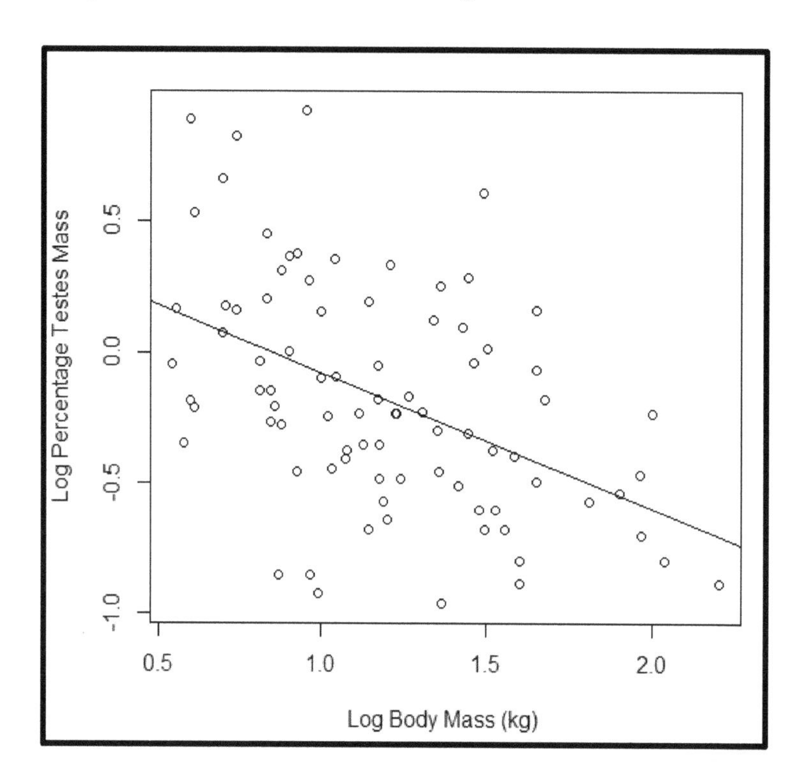

While the contents of your R CONSOLE window should look like this:

```
> cor.test(micro_bat_data$log_body_mass, micro_bat_data$log_ptm, method="pearson")

        Pearson's product-moment correlation

data:  micro_bat_data$log_body_mass and micro_bat_data$log_ptm
t = -4.7917, df = 83, p-value = 7.154e-06
alternative hypothesis: true correlation is not equal to 0
95 percent confidence interval:
 -0.6173752 -0.2801722
sample estimates:
       cor
-0.4655005

> plot(micro_bat_data$log_body_mass, micro_bat_data$log_ptm, xlab="Log Body Mass (kg)",
> abline(lm(micro_bat_data$log_ptm~micro_bat_data$log_body_mass))
> |
```

If you examine the results of this analysis you will see that there is a significant correlation between log-transformed body mass and log-transformed percentage testes mass (p-value=7.154e-06 or 0.000007154). However, with correlation analyses, the significance of the relationship is not the only important metric. It is also important to look at the value of the correlation coefficient itself. This will tell you how strong the correlation is, and whether it is positive or negative. As outlined above, the potential value of a correlation coefficient ranges between -1 (for a perfect negative correlation) and 1 (for a perfect positive correlation). A middle value of 0 indicates no correlation between the variables. In this instance, the correlation coefficient (cor) is -0.465505. This means that the relationship found by this analysis is relatively strong (as the absolute value of the correlation coefficient is almost half way between 0 and 1) and negative. To report the results of a correlation test, you need to identify the test you used, the value of the correlation coefficient (cor), the sample size (n), and the resulting p-value (p). In all cases, the values need to be rounded to an appropriate number of figures (see Appendix III for details). For the above example, you could report it as follows:

When the relationship between log-transformed body mass and log-transformed percentage testes mass in mircochiropteran bats was analysed, it was found that there was a relatively strong and significant negative correlation between the two variables (Pearson Product Moment Correlation Coefficient: cor=-0.47; $p<0.001$*; n=84).*

When testing for correlations between two variables using the above workflow, you can apply either a Pearson Product Moment Correlation Coefficient or a Spearman's Rank Correlation Coefficient. Information on these two correlation tests, along with the requirements of each test and the modifications to the `cor.test` command required to run them, are provided in the table below. Information about additional arguments you can include in the `cor.test` command can be found at *www.rdocumentation.org/packages/stats/versions/3.6.1/topics/cor.test.*

Correlation Test	**Test Requirements And How To Conduct It In R**
Pearson Product Moment Correlation Coefficient	In order to calculate a Pearson Product Moment Correlation Coefficient, there needs to be a linear relationship between the variables being compared. In addition, they have to have a bivariate normal distribution. To calculate a Pearson Correlation Coefficient, include the argument `method="pearson"` in the `cor.test` command. For example, to calculate a Pearson Correlation Coefficient for variables called `log_body_mass` and `log_ptm` in a data set called `micro_bat_data`, the `cor.test` command would be: `cor.test(micro_bat_data$log_body_mass, micro_bat_data$log_ptm, method="pearson")`

Correlation Test	Test Requirements And How To Conduct It In R
Spearman's Rank Correlation Coefficient	In order to calculate a Spearman's Rank Correlation Coefficient, there does not need to be a linear relationship between the variables being compared nor do they have to have a bivariate normal distribution. However, it does require that the relationship between the variables is monotonic. To calculate a Spearman's Correlation Coefficient, include the argument `method="spearman"` in the `cor.test` command. For example, to calculate a Spearman's Correlation Coefficient for variables called `log_body_mass` and `log_ptm` in a data set called `micro_bat_data`, the `cor.test` command would be: `cor.test(micro_bat_data$log_body_mass, micro_bat_data$log_ptm, method="spearman")`

For the next part of this exercise, you will customise the above approach by investigating whether a similar negative correlation exists between log-transformed body mass and log-transformed percentage testes mass in a second group of bats, the macrochiropterans (which lack complex echolocation and are commonly referred to as fruit bats). This will allow you to explore how you would run a correlation analysis when one of the variables has a non-normal distribution. To do this, you will first need to import a new data set, contained in a file called `macro_bat_data.csv`, into R by modifying the `read.table` command in step 2 so that it looks like this (required modifications are highlighted in **bold**):

```
macro_bat_data <- read.table(file="macro_bat_data.csv",
        sep=",", as.is=FALSE, header=TRUE)
```

Once you have run this modified version of the `read.table` command, you need to check that the data have been imported correctly. This can be done by modifying the commands provided in step 3 so that they look like this:

```
names(macro_bat_data)
View(macro_bat_data)
```

Next, you need to check whether or not the relationship between log-transformed body mass and log-transformed percentage testes mass in this new data set is also linear. To do this, you need to modify the `plot` command in step 4 so that it looks like the code at the top of the next page.

```
plot(macro_bat_data$log_body_mass, macro_bat_data$log_ptm)
```

When you examine the resulting scatter plot you will see that there is a broadly linear relationship between these two variables in the macrochiropteran data set. The next step is to test both variables for normality. This can be done by modifying the `shapiro.test` commands provided in step 5 so that they look like this:

```
shapiro.test(macro_bat_data$log_body_mass)
shapiro.test(macro_bat_data$log_ptm)
```

Once you have run these tests for normality, you will see that while the distribution of the log-transformed body mass data is not significantly different from normal (p-value=0.2079), there is a significant difference for the log-transformed percentage testes mass data (p-value=0.009671). This means that it would be inappropriate to calculate a Pearson Product Moment Correlation Coefficient from these data. Instead, you will need to calculate a Spearman's Rank Correlation Coefficient. This can be done by modifying the `cor.test` command provided in step 6 so that it look like this:

```
cor.test(macro_bat_data$log_body_mass,
macro_bat_data$log_ptm, method="spearman")
```

When you run this command, you may see a warning message about ties in your data. If you do not have any tied values in your data set, as is the case here, you can safely ignore this warning. If you examine the results, you will see that there is a relatively weak negative relationship between the two variables (cor=-0.1744361) and that this relationship is non-significant (p=0.4603). This lack of a significant correlation can be reported as follows:

When the relationship between log-transformed body mass and log-transformed percentage testes mass in macrochiropteran bats was analysed, it was found that there was a relatively weak and non-significant negative correlation between the two variables (Spearman's Rank Correlation Coefficient: cor=-0.17; p=0.460; n=20).

To visualise this weak relationship between the variables in macrochiropteran bats, you can modify the commands from step 7 so that they look like the code provided at the top of the next page.

```
plot(macro_bat_data$log_body_mass, macro_bat_data$log_
ptm, xlab="Log Body Mass", ylab="Log Percentage Testes
Mass")
abline(lm(macro_bat_data$log_ptm~macro_bat_data$log_
body_mass))
```

This should produce a final scatter plot that looks like this:

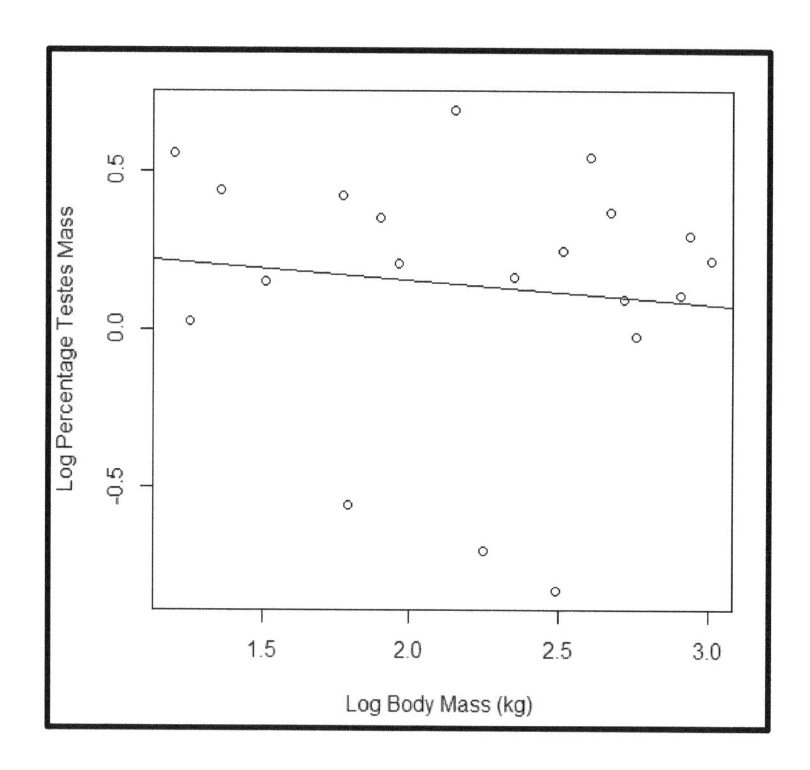

Once you have learned how to customise the workflow for conducting a correlation between two variables, you can test your ability to use it correctly by analysing the relationship between log-transformed body mass and log-transformed percentage testes mass in a different taxonomic group, the cetaceans. This will help you understand what to do when the preliminarily exploration of your data suggests there is a non-linear relationship between the variables being examined. To do this, you will first need to import the data set contained in a file called `cetacean_data.csv` by modifying the `read.table` command in step 2 so that it looks like this:

```
cetacean_data <- read.table(file="cetacean_data.csv",
sep=",", as.is=FALSE, header=TRUE)
```

After you have run this modified version of the `read.table` command, you need to check that the data have been imported correctly. This can be done by modifying the commands provided in step 3 so that they look like this:

```
names(cetacean_data)
View(cetacean_data)
```

Next, you need to check whether or not the relationship between log-transformed body mass and log-transformed percentage testes mass in this new data set is linear. To do this, you need to modify the `plot` command in step 4 so that it looks like this:

```
plot(cetacean_data$log_body_mass, cetacean_data$log_ptm)
```

When you examine the resulting scatter plot (see image below) you will see when you try to place a pen or pencil through the middle of the data, there are clearly more points above it for the species with the largest and smallest body masses, and more below in for intermediate body masses.

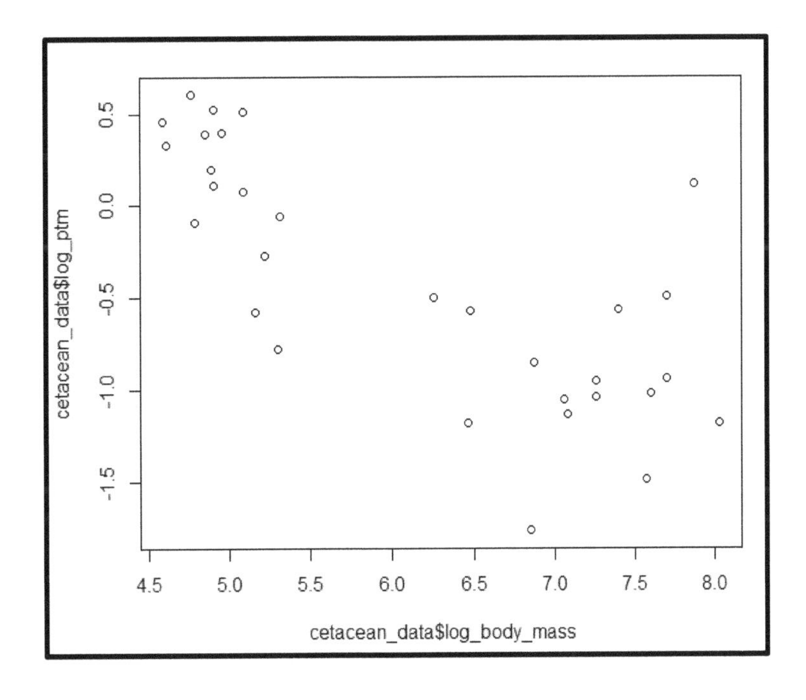

This is evidence that the relationships between these two variables in cetaceans is non-linear, meaning it would be inappropriate to calculate a Pearson Product Moment Correlation Coefficient from these data. Instead, you would need to calculate a Spearman's Rank Correlation Coefficient (as you have already ruled out the possibility of calculating a Pearson Correlation Coefficient, there is no need to apply any tests for normality before you do this). This can be done by modifying the `cor.test` command provided in step 6 so that it look like this:

```
cor.test(cetacean_data$log_body_mass, cetacean_data$log_
                    ptm, method="spearman")
```

When you run this command, you will see that there is a very strong negative relationship between the two variables (cor=-0.7164751) and that this relationship is significant (p=5.81e-06 or p=0.00000581). This relationship can be reported a follows:

When the relationship between log-transformed body mass and log-transformed percentage testes mass in cetaceans was analysed, it was found that there was a very strong and significant negative correlation between the two variables (Spearman's Rank Correlation Coefficient: cor=-0.72; p<0.001; n=31).

To visualise this relationship, you can modify the `plot` command from step 7 so that it looks like this:

```
plot(cetacean_data$log_body_mass, cetacean_data$log_ptm,
xlab="Log Body Mass", ylab="Log Percentage Testes Mass")
```

This should produce a final scatter plot that looks like the image at the top of the next page (**NOTE:** Since the relationship between the two variables is non-linear, it would be inappropriate to add a linear line of best fit to this graph).

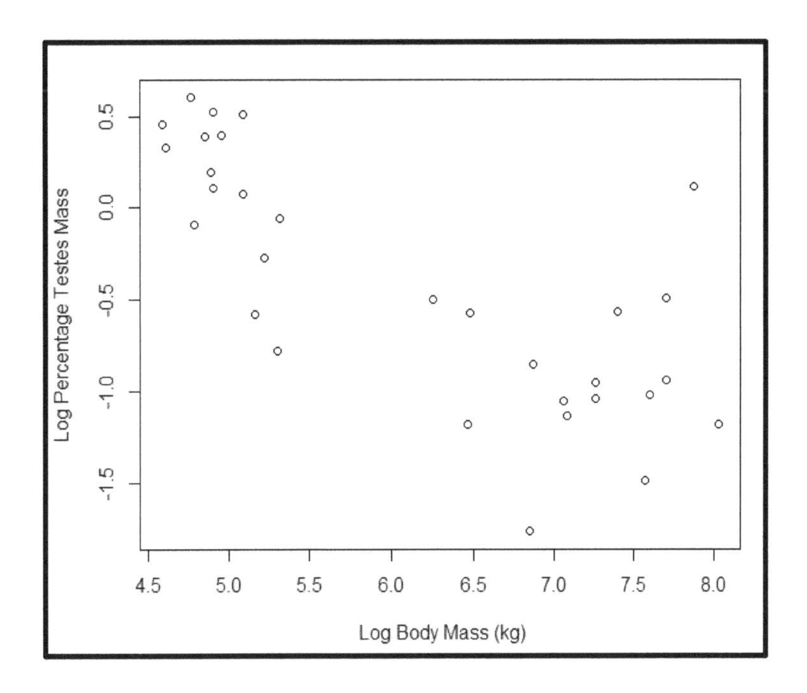

EXERCISE 5.2: HOW TO TEST FOR A RELATIONSHIP BETWEEN TWO VARIABLES USING LINEAR REGRESSION:

While a correlation simply tests whether there is a relationship between two variables, a regression allows you to test whether there is any evidence that changes in the value of one variable are dependent on changes in the other, and to define what this relationship is using a mathematical formula. This means that you can use a regression analysis to predict how the value of the response or dependent variable will change across a range of different values for the explanatory or independent variable, even if you have not measured this relationship at every point within it. As a result, regression analyses are generally used in biology when you wish to understand and predict how changes in one variable are driven by variations in another. Thus, one of the first decisions you will have to make before you can carry out a linear regression is which of the variables you are analysing is the response or dependent variable and which is the explanatory or independent variable. This is usually determined by the hypothesis you wish to test with your regression analysis. For example, for the bat data you will use in the first part of this exercise, the explanatory hypothesis might be that the level of proportional investment in testes mass is determined by the body mass. Based on this hypothesis, log-transformed percentage testes mass would be considered the response or dependent variable and log-transformed body mass would be considered the explanatory or independent variable.

Despite the large number of possible ways to carry out a regression analyses, most of them require you to complete the same basic steps, and for the data to have the same structure needed to calculate a correlation coefficient (see Exercise 5.1). To illustrate these steps, in the first part of this exercise you will conduct a simple linear regression analysis. This linear regression will be conducted on the data on the R object called `micro_bat_data` created Exercise 5.1. This means that you will have to have completed that exercise before you can start working through the following flow diagram (**NOTE:** If you wish to apply this workflow to another data set, you would need to conduct steps 1 to 4 from the flow diagram in Exercise 5.1 before you can start working through this one):

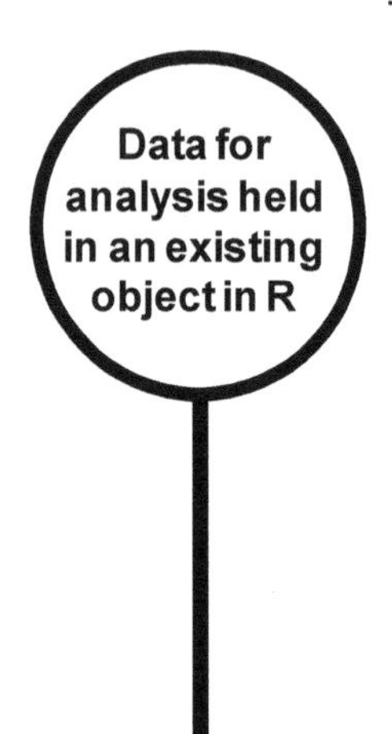
Data for analysis held in an existing object in R

For this example, the data set you will use is held in the R object called `micro_bat_data` created as part of Exercise 5.1.

Once you have decided which of your variables will serve as the dependent or response variable and which will serve as the independent or explanatory variable, you can create a linear regression model. While this can be done in a number of ways, for this example you will use the `lm` command to create it. To do this, enter the following commands into R:

```
micro_bat_data <- micro_bat_data[order
    (micro_bat_data$log_body_mass),]
bat_lm <- lm(log_ptm~log_body_mass, data=
    micro_bat_data)
```

1. Conduct a regression analysis to model the relationship between your response or dependent variable and your explanatory or independent variable

This code has to be entered exactly as it is written here or it will not work. If you wish to use the copy-and-paste approach for entering these commands, copy the text directly below CODE BLOCK 107 in the document R_CODE_BASIC_STATS_WORKBOOK.DOC and paste it into R.

The first part of this code sorts the `micro_bat_data` data set based on the values of your selected explanatory or independent variable (in this case, `log_body_mass`) using the `order` command. **NOTE:** if you don't do this, you will have problems comparing your residuals to your explanatory or independent variable in step 4. The second part of this code creates a linear regression model using the `lm` command on these sorted data. This model uses the data in the column called `log_ptm` from the R object called `micro_bat_data` as the response or dependent variable, and the data in the column called `log_body_mass` as the explanatory or independent variable. This model is stored in a new R object called `bat_lm`.

After you have created your linear regression model, you need to check if there are any problems with it that may invalidate its findings. This is done by checking the residuals of the model (this is the difference between the actual value for each data point and the one estimated by the model). For a linear regression model to be valid, these residuals need to have a normal distribution and not have any consistent biases when plotted against the independent or explanatory variable included in the analysis. To extract these residual values from your linear regression model, you will use the `resid` command. To do this for the model created in step 1 (called `bat_lm`), enter the following code into R:

2. Extract the residuals for your linear regression model

```
bat_lm_resid <- resid(bat_lm)
```

This is CODE BLOCK 108 in the document R_CODE_BASIC_STATS_WORKBOOK.DOC. This command creates a new R object called `bat_lm_resid` which contains the residual value for each data point in the data set used to create your linear regression model (in this case, `micro_bat_data`).

In order to check whether there are any unwanted patterns within the residuals from your linear regression model, you will need to do two things. The first is to assess whether they have a normal distribution, and the second is to plot their values against the values of the independent or explanatory variable included in the model.

3. Check whether the distribution of your residuals are normal

To assess whether your residuals have a normal distribution, you can conduct an appropriate normality test (see Exercise 3.1). For this example, you will apply a Shapiro-Wilk test to them by entering the following command into R:

```
shapiro.test(bat_lm_resid)
```

This is CODE BLOCK 109 in the document R_CODE_BASIC_STATS_WORKBOOK.DOC. When you run this command for the bat data being analysed in this example, you will find that the distribution of the residuals is not significantly different from normal (p-value=0.2087).

To examine the relationship between the independent or explanatory variable (in this case, `log_body_mass`) and the residual values from your linear regression model, you need to create a scatter plot of the relationship and assess whether it is significant. To do this, enter the following commands into R:

```
plot(bat_lm_resid~micro_bat_data$log_
body_mass)
abline(lm(bat_lm_resid~micro_bat_data$log_
body_mass))
bat_lm_resid_model <- lm(bat_lm_resid~
micro_bat_data$log_body_mass)
summary(bat_lm_resid_model)
```

This is CODE BLOCK 110 in the document R_CODE_BASIC_STATS_WORKBOOK.DOC. When you run these commands for the data being used in this example, you will find there is no evidence of any type of pattern between the residuals and the independent or explanatory variable, and the relationship between them is not significant (p-value=1).

4. Check the residuals for patterns that violate the assumptions of linear regression modelling

You may also wish to add lines to your scatter plot that represent the 95% confidence intervals of the relationship between the residuals and the independent or explanatory variable. To do this, enter the following commands into R:

```
bat_lm_resid_ci <- predict(bat_lm_resid_
model, interval="confidence")
bat_lm_resid_ci <- as.data.frame(bat_
lm_resid_ci)
lines(loess(bat_lm_resid_ci$lwr~micro_bat_
data$log_body_mass), lty="dashed")
lines(loess(bat_lm_rsid_ci$upr~micro_bat_
data$log_body_mass), lty="dashed")
```

This is CODE BLOCK 111 in the document R_CODE_BASIC_STATS_WORKBOOK.DOC.

Once you have checked that there are no problems with the residuals of the linear regression model you created in step 1 (in this case, `bat_lm`), you can examine its results. This is done using the `summary` command. To do this, enter the following code into R:

```
summary(bat_lm)
```

5. Examine the results of your linear regression model

This is CODE BLOCK 112 in the document R_CODE_BASIC_STATS_WORKBOOK.DOC. This command will bring up the details of your linear regression model, including information about whether the relationship between the variables is significant, its strength, and the exact form this relationship takes.

6. Plot a graph of the relationship between the two variables, including a line of best fit and 95% confidence intervals (if appropriate)

Finally, you can produce a scatter plot to visualise the relationship your model has identified between the dependent or response variable and the independent or explanatory variable. To do this for the model created in this example, enter the following command into R:

```
plot(micro_bat_data$log_body_mass, micro_
bat_data$log_ptm, xlab="Log Body Mass",
ylab="Log Percentage Testes Mass")
```

This is CODE BLOCK 113 in the document R_CODE_BASIC_STATS_WORKBOOK.DOC. This command creates a scatter plot from the `log_body_mass` column and the `log_ptm` column in the R object called `micro_bat_data` (created in step 2 of Exercise 5.1) and adds labels to the X and Y axes.

Next, if you have found that there is a significant linear relationship between the two variables, you can add a line representing this relationship to your graph. To do this for the model created in this example, enter the following command into R:

```
abline(lm(micro_bat_data$log_ptm~micro_
bat_data$log_body_mass))
```

This is CODE BLOCK 114 in the document R_CODE_BASIC_STATS_WORKBOOK.DOC

You may also wish to add lines to your scatter plot that represent the 95% confidence intervals of the modelled relationship. To do this, enter the following commands into R:

```
bat_lm_ci <- predict(bat_lm, interval=
"confidence")
bat_lm_ci <- as.data.frame(bat_lm_ci)
lines(loess(bat_lm_ci$lwr~micro_bat_data$log
_ body_mass), lty="dashed")
lines(loess(bat_lm_ci$upr~micro_bat_data$
log_body_mass), lty="dashed")
```

This is CODE BLOCK 115 in the document R_CODE_BASIC_STATS_WORKBOOK.DOC. This code uses the `predict` command to create a table (called `bat_lm_ci` in this example) with the upper (`upw`) and lower (`lwr`) confidence intervals in it. These data can then be used to generate lines representing the upper and lower confidence intervals using the `lines` command. This is done by plotting them against the data in the `log_body_mass` column of the R object called `micro_bat_data`.

Linear regression analysis conducted

The scatter plot of the relationship between the residual values from your regression analysis and the independent or explanatory variable (in this case, log-transformed body mass) created in step 4 should look like this (**NOTE:** On this graph, the solid line represents the modelled relationship and the dotted lines represent its 95% confidence intervals):

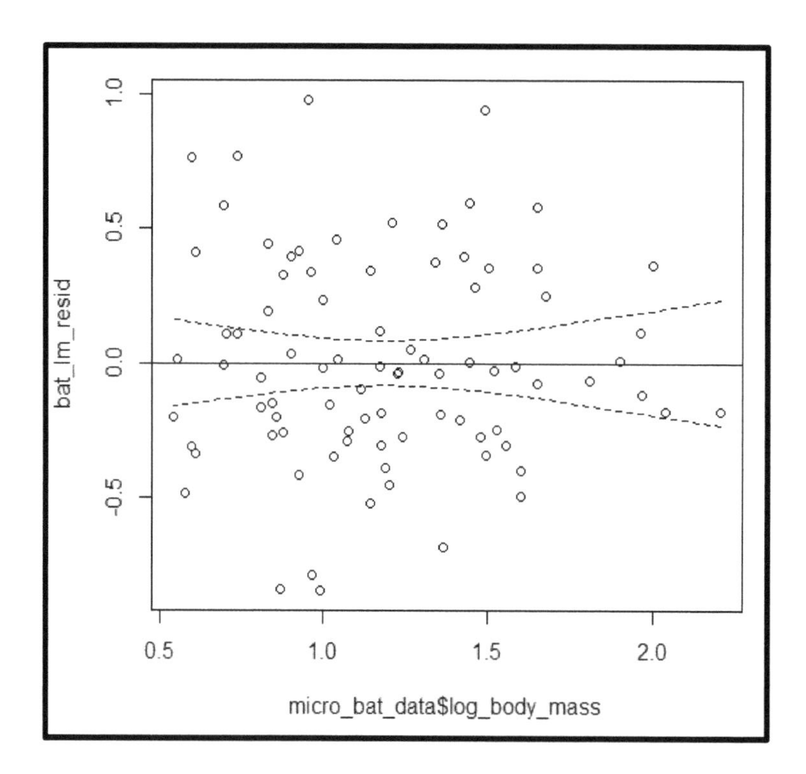

The shape of the line representing the relationship between the independent or explanatory variable and the residuals from the model, as well as the confidence intervals around it, and the lack of a significant relationship between them (see step 4 of the above flow diagram) confirm that there are no unacceptable patterns in the residuals of this model. This means that the assumptions of linear regression analysis are not violated for this particular model, and its results are likely to be valid.

The results of the linear regression analysis itself, which were brought up by the command `summary(bat_lm)` in step 5, should look like the image at the top of the next page (if this is no longer visible in your R CONSOLE window, you can re-enter this command and it will appear again).

```
> summary(bat_lm)

Call:
lm(formula = log_ptm ~ log_body_mass, data = micro_bat_data)

Residuals:
     Min       1Q   Median       3Q      Max 
-0.84569 -0.26103 -0.03566  0.32645  0.97913 

Coefficients:
              Estimate Std. Error t value Pr(>|t|)    
(Intercept)     0.4406     0.1356   3.250  0.00167 ** 
log_body_mass  -0.5203     0.1086  -4.792 7.15e-06 ***
---
Signif. codes:  0 '***' 0.001 '**' 0.01 '*' 0.05 '.' 0.1 ' ' 1

Residual standard error: 0.3857 on 83 degrees of freedom
Multiple R-squared:  0.2167,    Adjusted R-squared:  0.2073 
F-statistic: 22.96 on 1 and 83 DF,  p-value: 7.154e-06
```

From this summary, you can see that there is a significant relationship between log-transformed body mass and log-transformed percentage testes mass (p-value=7.15e-06 or 0.00000715) in microchiropteran bats, and that this relationship is moderately strong (Adjusted R-squared=0.2073).

The scatter plot of the relationship identified by the regression analysis between log-transformed body mass and log-transformed percentage testes mass created in step 6 should look like the image below. As before, the solid line on this graph represents the modelled relationship and the dotted lines represent its 95% confidence intervals.

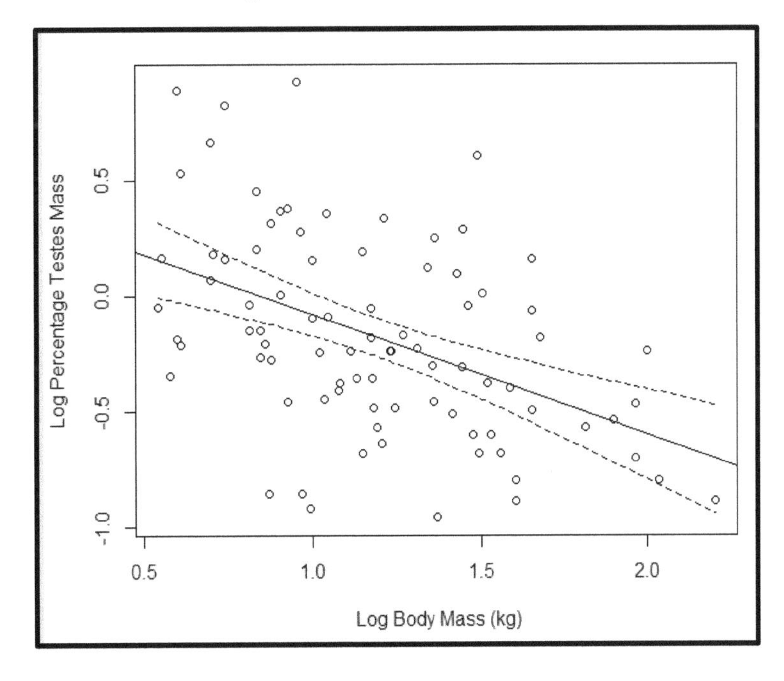

This shows the negative relationship between log-transformed body mass and log-transformed percentage testes mass in microchiropteran bats, meaning that males of larger microchiropteran bats invest proportionately less in their reproductive tissues than males of smaller species. If you look back at the results for this analysis, you will see that the line representing this relationship has a gradient or slope of -0.5203 and an intercept with the Y axis of 0.4406 (these values can be found in the Estimate column in the coefficients table in the summary of the results). From this, you can generate the equation used to draw the line representing this relationship based on the standard linear equation structure of Y=bX+c, where b is the slope and c is the intercept with the Y axis. For the above linear regression analysis, this equation would be:

log percentage testes mass = -0.5203 * log body mass + 0.4406

These findings can be reported as follows:

*There was a significant and moderately strong negative relationship between the log-transformed body mass of male microchiropteran bat species and their log-transformed percentage testes mass (linear regression: log-transformed percentage testes mass = -0.520 * log-transformed body mass + 0.441;* R^2=0.207; $t_{1,83}$=-4.8; p<0.001; n=84). *This means that males of larger species invest proportionately less in their reproductive tissues than males of smaller species.*

If you wish to learn about the additional arguments you can include in the `lm` command when conducting a linear regression analysis, you can find this information at *www.rdocumentation.org/packages/stats/versions/3.6.1/topics/lm.*

As mentioned in the introduction to this exercise, there are many different ways to conduct regression analysis. However, one of the most commonly used regression frameworks in biological studies is the GLM one. If you have completed Exercise 4.4 in this workbook, you will already have used this framework as an alternative way to conduct an ANOVA. Within regression analyses, this same framework can be used not only to conduct linear regressions, but also to conduct general linear modelling (which are linear regression models that include independent or explanatory variables that are factors as well as continuous variables), generalised linear modelling (which also allow you to conduct regression analyses on data with non-normal distributions), and generalised linear mixed modelling (which allows you to take the effect of random factor variables into account).

To use the GLM framework for a linear regression, you would simply replace the `lm` command in step 1 of the above flow diagram with the `glm` command. To give you experience with doing this, you can repeat the above analysis for the relationship between log-transformed body mass and log-transformed percentage testes mass in microchiropteran bats using the `glm` command instead of the `lm` command. You will also need to include the `family=gaussian` argument in the command so that it uses a normal distribution. To do this, modify the R command in step 1 so that it looks like this (required modifications are highlighted in **bold**):

```
bat_glm <- glm(log_ptm~log_body_mass, data=micro_
                bat_data, family=gaussian)
```

To complete the remaining steps in this workflow using the `glm` command, you will need to modify the commands by replacing `lm` with `glm` throughout, both in the names of R objects, and in the commands themselves (**NOTE**: You will not be able to generate confidence intervals to plot on your graphs with the code provided in this workflow when using the GLM framework). If you have done this successfully, the summary for your linear regression analysis, which can be called up using the `summary(bat_glm)` command, should look like this:

```
> summary(bat_glm)

Call:
glm(formula = log_ptm ~ log_body_mass, family = gaussian, data = micro_bat_data)

Deviance Residuals:
     Min        1Q    Median        3Q       Max
-0.84569  -0.26103  -0.03566   0.32645   0.97913

Coefficients:
              Estimate Std. Error t value Pr(>|t|)
(Intercept)     0.4406     0.1356   3.250  0.00167 **
log_body_mass  -0.5203     0.1086  -4.792 7.15e-06 ***
---
Signif. codes:  0 '***' 0.001 '**' 0.01 '*' 0.05 '.' 0.1 ' ' 1

(Dispersion parameter for gaussian family taken to be 0.148762)

    Null deviance: 15.763  on 84  degrees of freedom
Residual deviance: 12.347  on 83  degrees of freedom
AIC: 83.236

Number of Fisher Scoring iterations: 2

> |
```

From this summary, you can see that this analysis found the same significant relationship that was found when using the `lm` command. If you wish to learn about the additional arguments you can include in the `glm` command when using it to conduct linear regression, you can find out more at *www.rdocumentation.org/packages/stats/versions/3.6.1/topics/glm.*

In order to gain further experience in using this workflow to conduct a linear regression analysis in R, you can analyse the relationship between log-transformed percentage testes mass and log-transformed body mass in the cetacean data set used at the end of Exercise 5.1 (this should be in an R object called `cetacean_data`). When you do this using `lm` command (rather than the `glm` command), you will find that the graph of the relationship between the residuals from the analysis and log-transformed body mass looks like this:

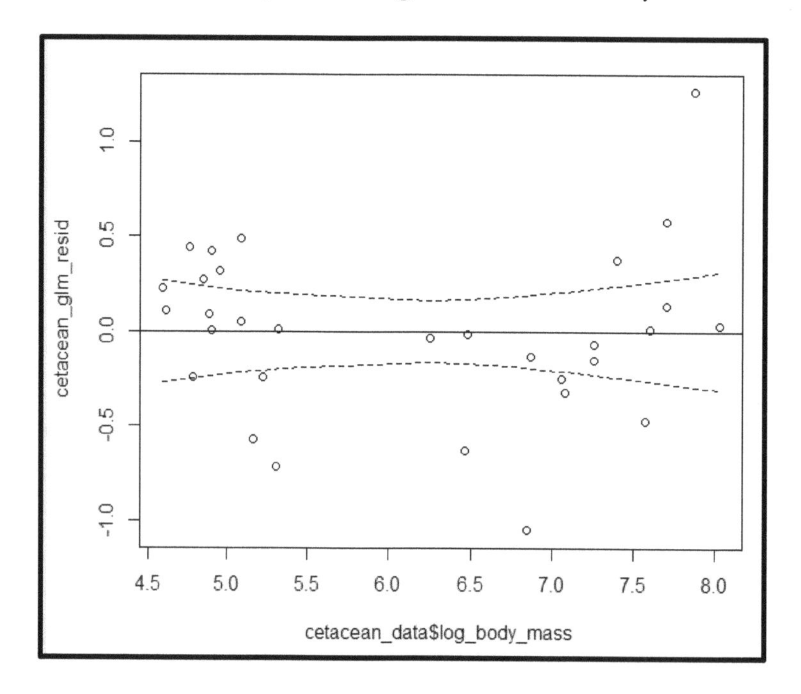

If you examine this graph, you will see that there is clearly a pattern within it that invalidates the assumptions of linear regression. This is because a disproportionate number of the residual values for the largest and smallest cetacean species are above the line of best fit, while all the residual values for the cetacean species with intermediate body masses are below it. This indicates that the relationship between these variables in cetaceans is non-linear and, as a result, that it is not appropriate to analyse these data with a linear regression framework. Instead, you would need to analyse them with an alterative framework that allows for non-linear relationships, such as generalised additive modelling (GAM).

--- Appendix I ---

A Simple Strategy For Working Out How To Do A Specific Task In R

While this workbook provides examples of how to do a range of tasks biologists commonly have to do, it is likely that you will also wish to complete either variations on these tasks or other similar tasks in R. This raises the question of how you can work out how to do them. While this may seem daunting at first, once you start doing it, it rarely turns out to be as difficult as it initially seems. The key to being able to do this is to break down the overall task into individual steps and then work out how to do them. One of the easiest ways to do this is to use the same flow diagram-based approach used for the exercises in this workbook. To create your own flow diagram for a new task, start by defining what you wish to be able to achieve with it. This is represented by the diamond at the end of each flow diagram. Next, identify your starting point. This is usually the data set or data sets you wish to work with. These are represented by circles at the top of the flow diagrams. Once you have your start and end points, you need to identify the major steps that will take you from one to the other. In the flow diagrams used in this workbook, these are the numbered boxes on the left hand side. Finally, once you have identified these major steps you can then work out exactly what you need to do to complete each one. This information is recorded in the text on the right hand side of the flow diagrams.

The above description makes it seem like this is a simple linear process, but it is unlikely that it will be. As you fill out your flow diagram, you should also be testing out the workflow it represents by running the R code you intend to use for each command. During this testing, you may find that some sections you thought would be a single step need to be broken down into two or more separate steps. Similarly, you may find that the way you thought you could do something will not work, and that you need to replace a specific step with a new one. As a result, as you are working out how to do a specific task, your flow diagram will be a living document that will be changing all the time as you figure out how to do its individual parts. However, once you have finished it, you will have a complete workflow that will allow you do to the same thing over and over again.

To show this process in operation, the following illustrations provide an example based on the development of the flow diagram for testing whether there is a difference in the body mass of male and female great tits from the start of Exercise 4.1. These images assume that the flow diagrams are being hand-drawn, as this is often the quickest and most flexible way to do it, especially at the start when you are likely to have to make many changes to it. However, once you have completed your flow diagram, you may wish to create a neater final copy using software such as PowerPoint or Word. The first step in this process is to define the start and end points, and use these to begin creating your flow diagram. At this point, it would look like this:

Next, you can identify the major steps that you think are required to get from the start point to the intended end point. In this case, five major steps can be identified. These are: 1. Set a working directory; 2. Import the required data into R; 3. Check the data have been imported correctly; 4. Create a box plot to explore the data; 5. Run a t-test on the data. Adding these to the above flow diagram gives you one that looks like this:

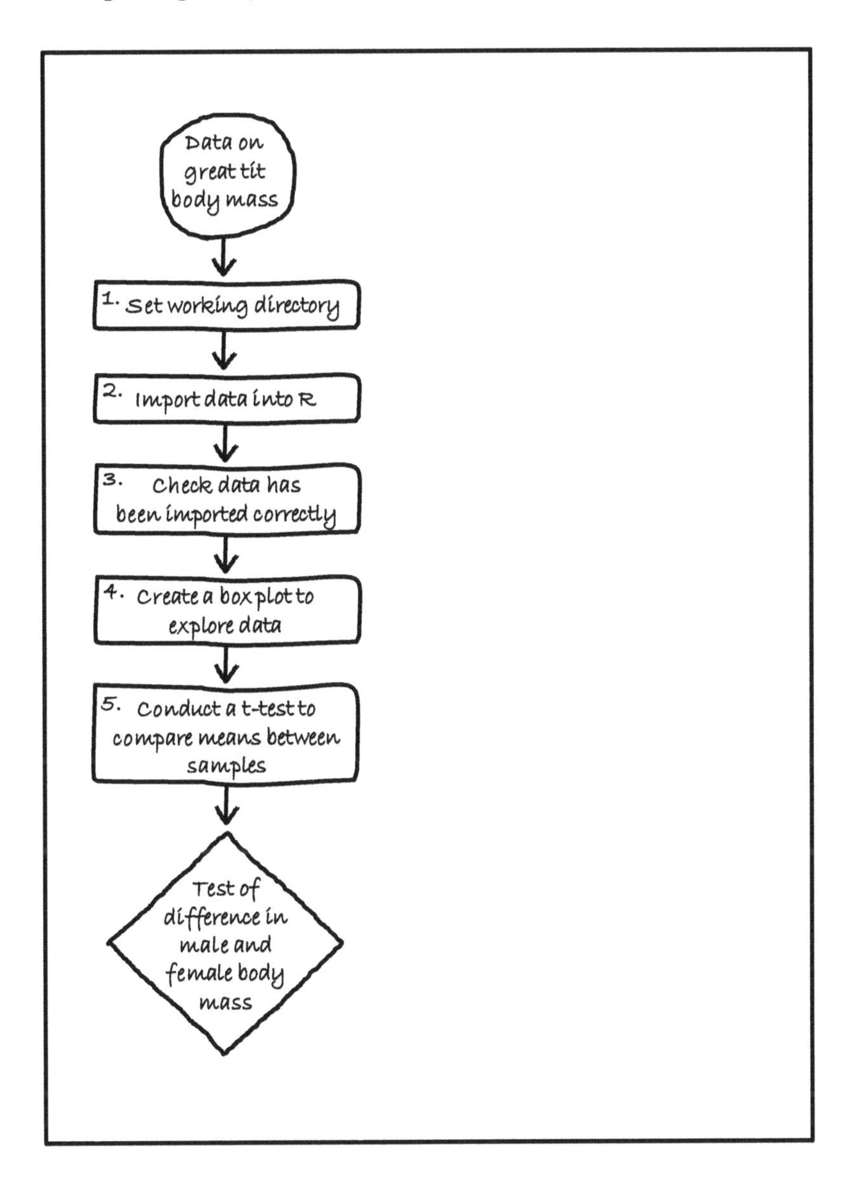

You can now fill in the details of the code required to complete each step. Often the best way to find out how to do each individual step is to use a search engine such as Google, or to ask a colleague if they know how to do it. However, the more experienced you become with running analyses in R, the more you will find that you can simply re-use individual blocks from other flow diagrams. For example, almost all workflows will include a data import step, meaning you can copy this step repeatedly into different flow diagrams. However, when you do this, you need to remember to update the exact details provided in the text on the right hand side to adapt it to each specific circumstance. Once you have added all the relevant information to it, your flow diagram will look like this:

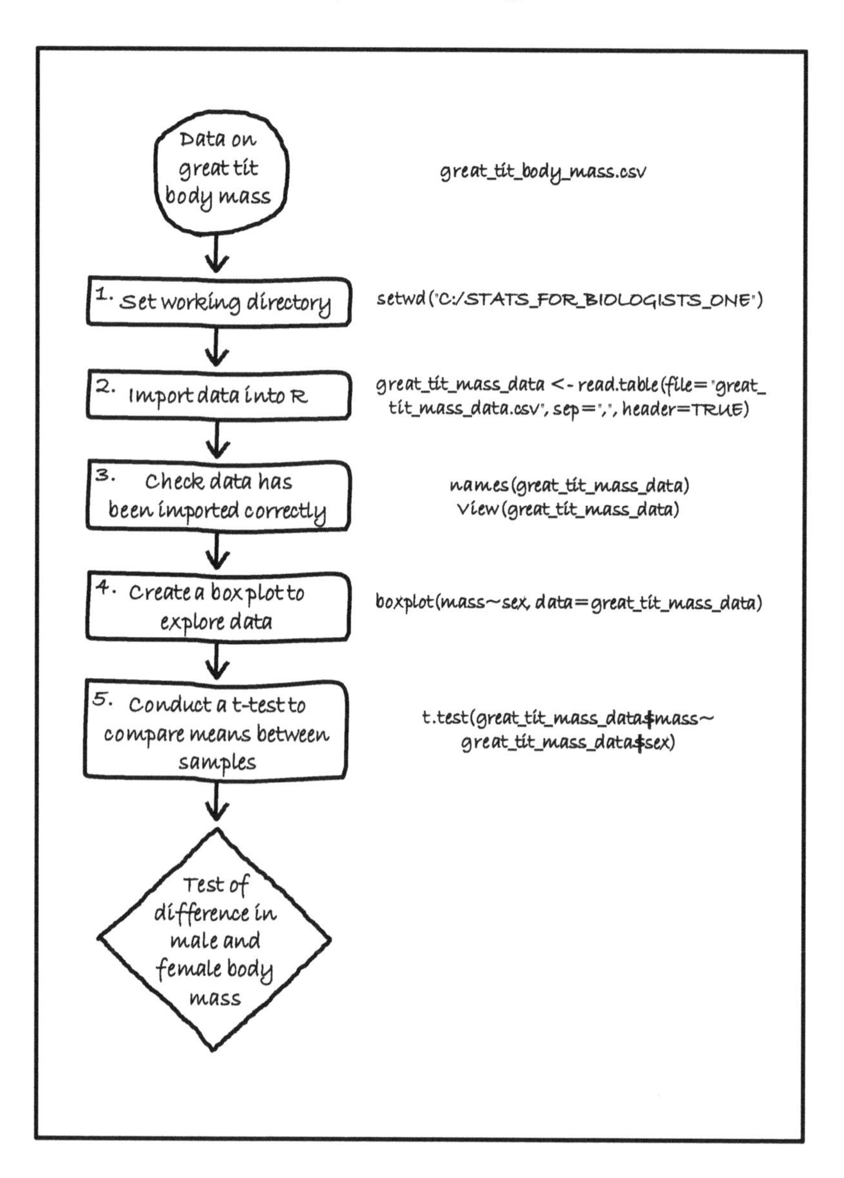

Now you have outlined your initial workflow using a flow diagram, you need to test it. When you do this, you will find that while steps 1 to 4 go according to plan, there is a problem with step 5. This is that a t-test requires that the data from each group are normally distributed, and you do not have a step in the current version of your workflow where this is checked. As a result, you would need to amend the workflow to include an extra step where the data are checked for normality. This would result in an updated flow diagram that looks like this:

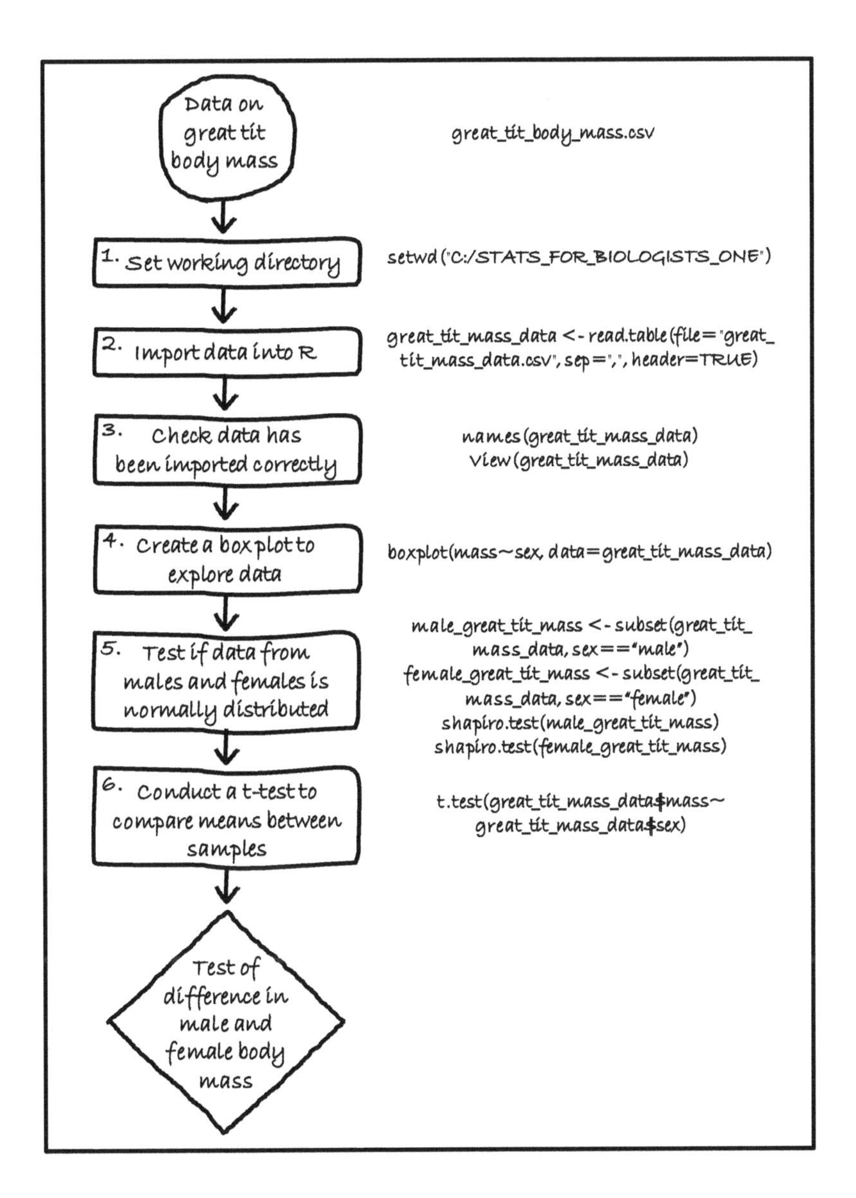

This new workflow can then be tested again to ensure that it now contains all the steps required to get from your starting point to your intended end point. For this example, it does, but for your own workflows this might not be the case. In particular, if you had found that your data were not normally distributed, you might have had to go back and add another step involving data transformation, or you might have decided that you need to change the test you use in step 6 from a t-test to a Mann-Whitney U test. When you have completed your flow diagram, and you are satisfied that it will allow you to achieved your intended goal, you are ready to implement it and to see what results it gives you.

After you have got the results of your analysis, it is always a good idea to create an archive version of the code you used to get them which can be re-load into R if you need to carry out the same or a similar analysis again in the future. Details of how to do this are provided in Appendix II.

--- Appendix II ---

How To Create An Annotated R Code Archive

Whenever you carry out a task in R, such as creating a graph or running a statistical analysis to help you test a specific hypothesis, it is important that you keep an accurate record of exactly what you did. There are three reasons for this. Firstly, it may be many weeks, months or even years before you need to do the same or a similar task again, and if you do not keep a detailed record of exactly what you did, it is likely that you will forget some or all of the steps involved. This means you would have re-learn how to the same task all over again, which is both annoying and time-consuming. If you keep a record of what you did previously, including copies of all the R commands you used, you can simply refer back to it, making it much easier to repeat the same task again at a later date.

Secondly, keeping an accurate record of what you have done will make it much easier when the time comes for you to write up your analysis as a thesis chapter or as a manuscript that you will submit for publication. This is because you will have a record of exactly what you did which you can refer back to. In addition, it is now common practice for academic journals to require you to provide a copy of all your R code as supplementary materials alongside your manuscript to allow a full assessment of whether you have conducted your analysis in an appropriate way. Again, if you create a record of what you did as part of any analyses, you will already had a copy of this information which you can submit, and this is much easier than having to re-create it at a later date.

Finally, by keeping an accurate record of what you do, you will be able to share this information with others, allowing you to contribute to increasing the skills base of a wider community, such as your research group or your field as a whole. It also means that if you are ever tasked with having to teach an analytical method you have used to others, you will have a ready-made set of instructions that you can give to them, rather than having to personally demonstrate it to each and every new person you are required to teach it to, saving you a lot of time. We know this from personal experience, and this workbook grew out of our own need to provide exactly this type of information to students and colleagues on a regular basis.

Creating a complete record for a specific task you have carried out in R is not simply a matter of saving the contents of your R CONSOLE as a workspace file or the contents of your SCRIPT EDITOR window as an R Script file (which will have the extension .R). This is because these windows will undoubtedly have pieces of code which did not work properly (for example, they may have had typos in them), pieces of code that turned out not to do quite what you wished them to do (for example, when you used the wrong command), and pieces of code that did not have the right settings for either required arguments or any additional arguments you included. Thus, in order to create an archived record of the exact method you used, you will need to edit down your code so that it only contains the final version of each command you used to run your analysis from start to finish. While this is difficult to do in the R CONSOLE window of RGUI, it is much easier to do in the SCRIPT EDITOR window of RStudio. As a result, we strongly recommend that you use RStudio as your user interface for running your analyses in R.

However, cleaning up your code is not the only step you need to carry out to create a complete archive of what you have done. You also need to annotate it so that anyone reading it (including yourself) can easily understand exactly what it was created to do and how it does it. While you can add annotations in any way you wish, we recommend using the format outlined below as this allows you to create a complete and easy-to-understand archive. There are three types of annotations in this format. These are:

1. **A Title And Summary:** These annotations appear at the very start of your R code archive and their function is to provide a record of what analysis the archived code is designed to carry out and what data it was written to work with. It should also contain information about when you conducted the analysis detailed in the archive file and your contact details (in case anyone has any questions about what exactly your code does).

2. **Section Headings That Break Your Code Into Individual Steps:** These annotations help you, and others, understand the aim of each individual section of code. These individually numbered steps are analogous to the individual blocks that make up the left hand side of the flow diagrams provided at the start of each exercise in this workbook.

Indeed, you may find it useful to actually create such flow diagrams for yourself to accompany your R code archives (see Appendix I for information on how to do this). However, simply numbering them is usually enough to help you remember what the major steps are in a specific analysis. If you do not break your code up into these individually numbered steps, you will find it much more difficult to work out which pieces of code refer to which major steps in the overall workflow, making it both harder to remember (and write up) what you did, and harder for you (or others) to adapt your code to conduct similar analyses on other data sets in the future.

3. **Comments That Explain What Each Section Of Code Does:** These comments are annotations that you add either before or after each individual R command (or block of R commands) that explain exactly what it does and what specific settings you used for the arguments you included in it. These comments are analogous to the detailed instructions provided on the right-hand side of the flow diagrams provided in this workbook.

In all cases, these annotations should start with hash tags (#). This is because R interprets anything that comes immediately after this symbol as being text that should be ignored when running the code it is associated with. This allows you to add whatever notes you wish to add without the risk of it causing the code itself not to run properly. To help differentiate between the different types of annotations, three hash tags should be used for the title and summary (###), two for section headings (##), and one for comments linked to specific pieces of code (#). After you have added these annotations, you will have created a version of your R code which can not only be easily re-loaded into R if you ever need to run the same, or a similar, analysis again, but that also contains all the information you need to remind yourself (or explain to others) exactly what each part of it does.

Once you have created the annotated version of your R code, you can then create an archive file from it. This can be done either by saving it as an R Script file (the format we recommend – see page 13 for details of how to do this) or as a simple text document.

To illustrate what an archived version of the R code for a specific workflow should look like, below is a fully annotated version of an R Script file for the t-test carried out by following the flow diagram from the start of Exercise 4.1 in this workbook. If you look back at this example, you will see that the numbered steps used in this archive file correspond to the major steps represented by the boxes on the left hand side of this flow diagram, while the annotations for the individual commands/blocks of commands correspond to the information contained in the text on the right hand side. Thus, using this structure for the archived version of your R code allows you to re-create this easy-to-understand flow-diagram-based structure within a text-only file format.

An Example Of The Contents Of An Annotated R Code Archive File

Title: A Comparison Of Mean Body Mass In Male and Female Great Tits

Summary: This R Script contains an archive of the R code used to compare the body mass of male and female great tits. It was carried out by [*insert your name here*] on [*insert date here*]. This was based on the data set `great_tit_mass_data.csv` collected from a study site in central Scotland. For more information, contact [*insert your contact details here*].

Step 1: Set The Working Directory For The Analysis In R:

```
setwd("C:/STATS_FOR_BIOLOGISTS_ONE")
```

This command sets the working directory for this analysis to C:/STATS_FOR_BIOLOGISTS_ONE.

Step 2: Import The Required Data Into R:

```
great_tit_mass_data <- read.table(file="great_tit_mass_
data.csv, sep=",",as.is=FALSE, header=TRUE)
```

This command imports a file called `great_tit_mass_data.csv` from the above working directory into R and uses it to create a new R object called `great_tit_mass_data`.

Step 3: Check The Data Have Been Imported Into R Correctly:

```
names(great_tit_mass_data)
```

This command returns the names present in the R object created using the `read.table` command from step 2, allowing you to verify that the required data have been imported correctly

```
View(great_tit_mass_data)
```

This command opens up a data viewer window where you can examine the contents of the R object created using the above `read.table` command, allowing you to verify that the required data have been imported correctly in a second way.

Step 4: Create A Box Plot To Provide An Initial Assessment Of The Distribution Of The Data:

```
boxplot(mass~sex, data=great_tit_mass_data)
```

This command creates a box plot that will show the range of values in the `mass` column for each group listed in the `sex` column in the R object called `great_tit_mass_data` created in step 2.

Step 5: Assess Whether Or Not The Distribution Of Data For Male And Female Great Tits Is Significantly Different From Normal

```
male_great_tit_mass <- subset(great_tit_mass_data,
sex=="male")
female_great_tit_mass <- subset(great_tit_mass_data,
sex=="female")
```

This pair of commands creates separate R objects for the data from male and female great tits. This is required to allow separate Shapiro-Wilk normality tests to be run on the mass data for each sex.

```
shapiro.test(male_great_tit_mass_data$mass)
```

This command runs a Shapiro-Wilk test to assess whether the distribution of the mass data for male great tits differs significantly from normal. It does not differ significantly from a normal distribution (p=0.5607), so these data can be used in a parametric test.

```
shapiro.test(female_great_tit_mass_data$mass)
```

This command runs a Shapiro-Wilk test to assess whether the distribution of the mass data for female great tits differs significantly from normal. It does not differ significantly from a normal distribution (p=0.4676), so these data can be used in a parametric test.

Step 6. Conduct An Appropriate Test To Assess Whether There Is A Significant Differences In Body Mass Between Male And Female Great Tits:

```
t.test(great_tit_mass_data$mass~great_tit_mass_data$sex)
```

This command runs a t-test that compares the mean values for the data in the `mass` column based on the categories in the `sex` column of the R object `great_tit_mass_data` created in step 2. This test was selected based on the results of the normality tests conducted in step 5. This found that there was a significant difference in mean body mass between the two sexes (t-test: t =-5.6; n for males=50; n for females=50; d.f. =84.7; p<0.001). This means that, on average, male great tits are significantly heavier than females.

--- Appendix III ---

How To Round Numbers When Reporting The Results Of Statistical Analyses

The results of statistical analyses in R are often provided to a high level of precision (indicated by number of decimal places). This may seem like a good thing, but the level of precision provided is often so great as to be biologically meaningless. As a result, when reporting the results of statistical analyses, you need to round the numbers provided by R to a more meaningful level of precision. Unfortunately, there are no hard and fast rules for how best to do this. Instead, there are a series of more flexible 'rules-of-thumb' that vary depending on exactly what numbers you are reporting, how big they are, and the purpose you are using them for. In particular, many academic journals have their own specific rules for rounding numbers when reporting the results of statistical analyses.

Below, you will find a set of general guidelines that will be suitable for most purposes, and that will require the minimum amount of modifications to adapt them to almost all possible circumstances. However, in all cases, rounding should only ever be applied to the final result and not at any intermediate stages during an analysis.

Number	Guideline for Rounding
Central Values (Means and Medians)	If central values are greater than 1, they should be rounded to one decimal place. If they are less than one, they should be rounded to two significant figures. For example, a mean of 10.00009 should be rounded to 10.0, while a mean of 0.005736 should be rounded to 0.0057.
Measures of Variance (e.g. Standard Deviations, Standard Errors, and Interquartile Ranges)	If measures of variance are greater than 1, they should be rounded to one decimal place. If they are less than one, they should be rounded to two significant figures. For example, a standard deviation of 2.65243 should be rounded to 2.7, while a standard error of 0.0004627 should be rounded to 0.00046.
Degrees of Freedom (d.f.)	When the degrees of freedom is a whole number, it should be reported without rounding. If it includes a decimal fraction, it should be rounded to one decimal place. For example, a degree of freedom of 11 should not be rounded up or down, while one of 11.667 should be rounded to 11.7.

Number	**Guideline for Rounding**
Sample Size (n)	Sample sizes should never be rounded up or down.
Test Statistics (e.g. D, U, W and X^2)	If a test statistic is greater than one, it should be rounded to one decimal place. If it is less than one, it should be rounded to two significant figures. For example, a test statistic of 238.382 should be rounded to 238.4, while a one of 0.73663 should be rounded to 0.74.
Correlation Coefficients (cor)	Correlation coefficients should always be rounded to two decimal places. For example, a correlation coefficient of 0.423 should be rounded to 0.42.
R^2	If expressed as a percentage, R^2 (a measure of the strength of relationships in regression analyses) should be rounded to one decimal place. For example, an R^2 of 24.672% should be rounded to 24.7%. If expressed as a proportion, R^2 should be rounded to two decimal places. For example, an R^2 of 0.3574 should be rounded to 0.36.
P-value or Probability (p)	Probability values (p) should be rounded to three decimal places. For example, a p-value of 0.6728 should be rounded to 0.673, while a one of 0.001236 should be rounded to 0.001. Values below 0.001 should be reported as $p<0.001$.

In order to decide whether a number should be rounded up or rounded down, examine the digits that will be dropped. If there are more than two of them and if, when reading from left to right, they start with a 4, round the number down. If they start with a 5, then round it up. For example, a mean of 10.2376 should be rounded to 10.2, while a correlation coefficient of 0.2367 should be rounded to 0.24. If the digit that is to be dropped is exactly 5 (followed by no other digits), it should be rounded to the nearest even number. For example, an R^2 of 13.45% should be rounded down to 13.4%, while a p-value of 0.0035 should be rounded up to 0.004.

Index Of R Commands And Arguments Used In This Workbook

Commands are written in **bold**, while arguments are written in *italics*. Where the same piece of code is used both as command in its own right and as an argument in another command in this workbook, it is written in both **bold** and *italics*.

CPSIA information can be obtained
at www.ICGtesting.com
Printed in the USA
LVHW112308191121
703844LV00007B/471